D0881269

BRADFORD
COLLEGE
LIBRARY

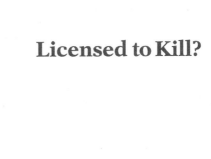

Licensed to Kill?

*Pitt Series in Policy and Institutional Studies*

Bert A. Rockman, *Editor*

# Joan Aron

# Licensed to Kill?

The Nuclear Regulatory

Commission and the

Shoreham Power Plant

University of Pittsburgh Press

HD
9698
.U54
S543
1997

Copyright © 1997, University of Pittsburgh Press

All rights reserved

Manufactured in the United States of America

Printed on acid-free paper

10   9   8   7   6   5   4   3   2   1

Library of Congress Cataloging-in-Publication Data

Aron, Joan B.

 Licensed to kill? : the Nuclear Regulatory Commission and the Shoreham

Power Plant / Joan Aron.

  p. cm. — (Pitt series in policy and institutional studies)

 Includes index.

 ISBN 0-8229-5649-7 (pbk. : acid-free paper). — ISBN 0-8229-4044-2

(cloth : acid-free paper)

 1. Shoreham Nuclear Power Station (N.Y.) 2. Nuclear power plants—Licenses—

New York (State) 3. Nuclear industry—Government policy—United States.

4. U.S. Nuclear Regulatory Commission. I. Title. II. Series.

HD9698.U54S543 1997

333.792'4'0979727—dc21         97-21163

                       CIP

A CIP catalog record for this book is available from the British Library.

BRADF

OCT – 6

BK. HSC   23.30

# Contents

# Acknowledgments

I wish to express my heartfelt thanks to a good friend and
collaborator, Leo Slaggie, deputy solicitor of the Nuclear
Regulatory Commission. This book was a cooperative
endeavor. It could not have been written without Leo's good
counsel, many suggestions, and generous assistance. He has
been involved with the project from its inception—as legal
and scientific advisor, critic, and editor. It has been an
enormously happy and helpful association. Leo's uncommon
intelligence, quick wit, and unfailing good humor have bright-
ened my e-mail and have made the manuscript far more lively
and interesting than it otherwise would have been.

# Abbreviations

| | |
|---|---|
| AEC | Atomic Energy Commission |
| ASLB | Atomic Safety and Licensing Board |
| ASLAB | Atomic Safety and Licensing Appeal Board |
| CEQ | Council on Environmental Quality |
| CP | construction permit |
| CPB | Consumer Protection Board |
| DOE | Department of Energy |
| DPC | Disaster Preparedness Commission |
| ECCS | emergency core-cooling system |
| EIS | Environmental Impact Statement |
| EPZ | "emergency planning zone" |
| FEMA | Federal Emergency Management Agency |
| GAO | General Accounting Office |
| GE | General Electric |
| IRS | Internal Revenue Service |
| Kemeny Report | President's Commission on the Accident at Three Mile Island |

| | |
|---|---|
| LHSG | Lloyd Harbor Study Group |
| LILCO | Long Island Lighting Company |
| LIPA | Long Island Power Authority |
| LOCA | loss-of-coolant accident |
| NEPA | National Environmental Policy Act of 1969 |
| NRC | Nuclear Regulatory Commission |
| NYPA | New York Power Authority |
| NYS | New York State |
| OL | operating license |
| POL | "possession only" license |
| PSC | Public Service Commission |
| RICO | Racketeer Influenced and Corrupt Organizations Act |
| School District | Shoreham–Wading River Central School District |
| SE2 | Scientists and Engineers for Secure Energy, Inc. |
| SOC | Shoreham Opponents Coalition |
| TMI | Three Mile Island |

# Foreword

*E. Leo Slaggie*

For the mid-Atlantic states Friday, July 19, 1991, was another sweltering day in what looked so far like the hottest year yet recorded. With temperatures up and down the coast hanging in the nineties, even in Maine, air conditioners fought to make the heat bearable, and electric utility spokesmen announced that the entire mid-Atlantic power pool was expected to reach a record usage level. Nuclear power provides a significant part of that electricity particularly in the heavily populated areas of the Northeast, where generating power by fossil fuel combustion costs more money and probably causes more serious environmental impacts than in most other parts of the United States. But on that day of record heat and power usage, one nuclear plant stood idle—a facility licensed to operate two years before, tested and physically ready to feed more than eight hundred megawatts of air pollution–free generating capacity into the power grid. The Shoreham Nuclear Power Station on Long Island, constructed at a cost of well over $5 billion, had not yet generated a single kilowatt, and almost everybody with any say in the matter seemed determined that Shoreham never would. For better or worse, they were about to prevail.

On that stifling July 19, the U.S. Nuclear Regulatory Commission (NRC), the federal agency responsible for licensing and regulating nuclear power plants, set its own uncomfortable kind of record: the commission approved the dismantling of a freshly licensed nuclear plant that was ready to operate, a plant that the commission itself had declared safe, needed, and environmentally desirable. After conducting years of contentious public hearings on the proposed licensing of Shoreham, after changing federal rules where necessary to make a license possible despite state and local noncooperation, and after defeating opponents in federal court, the NRC with equal flexibility and considerably greater speed now took back the Shoreham operating license it had granted in 1989. Once again, the commission was simply giving the Long Island Lighting Company (LILCO), the Shoreham licensee, what it wanted. What LILCO wanted now was to get out of the nuclear business, having reached an agreement with New York and Suffolk County not to operate the plant it had spent so much money and time building. The order of July 19 amended the operating license to "possession only," allowing LILCO to start taking Shoreham apart.

Not everyone agreed with the decision to write off Shoreham. The U.S. Department of Energy, the Council on Environmental Quality, the Shoreham-Wading school district, and an organization of pronuclear scientists had asked the NRC to hold off amending the license long enough at least to consider whether there were sensible alternatives to destroying this major non-fossil-fuel generating facility, but the commission said that was none of its business. The opponents of dismantling, supported by the U.S. Department of Justice, then asked the U.S. Court of Appeals to stay the amendment temporarily so that the court could review the case before Shoreham was made irretrievably inoperable. The judges declined to interfere, as did the U.S. Supreme Court. Thus, the most expensive nuclear power plant ever built was officially consigned to the scrap heap without ever producing any electricity.

Although the NRC disclaimed all authority or responsibility regarding the decision not to operate Shoreham, the decision was nevertheless in large part a governmental one. New York State and Suffolk County had settled their disputes with LILCO by offering various benefits and tax and rate relief in exchange for LILCO's agreement not to operate Shoreham and to transfer ownership of the plant to state authorities for $1 for eventual decommissioning. The state and county concern that motivated this agreement was their belief that the plant was unsafe and that if an accident happened emergency evacuation would be impossible, given the plant's Long Island location. With the support of the voters and ratepayers who would eventually be faced with the bill for this settlement, the state and county were willing to write off the multibillion-dollar Shoreham investment rather than accept the risks they perceived would come from operating the plant.

Shoreham appears to be the largest engineering project in all history that ever was completed and then abandoned without ever being used. Viewed from either side of the intense controversy over nuclear power, the decision to scrap Shoreham should raise concern for how well our governmental institutions are functioning. How could it happen that an enormously expensive power generation facility could be constructed and licensed with federal approval from start to finish, based on exhaustive safety review and analysis by a supposedly expert agency specifically created by Congress to perform such oversight, only to be discarded by state and local governments because of safety concerns? The NRC had found "reasonable assurance" that Shoreham could and would be operated safely, as the Atomic Energy Act requires. The commission said that nothing about the Long Island site would prevent an adequate response in the event of a radiological emergency. Thus the immediate cause of the Shoreham debacle was simply that the highest-ranking state and county officials did not believe the NRC, and neither, apparently, did a majority of the public affected by the proposed operation of the plant. If not for this distrust, Shoreham would be operating, and the resources that LILCO had over-liberally poured into its construction (ultimately to be paid by the public in some way or another) would not have gone totally to waste.

But would it be operating safely? One cannot look for the "mistakes" that brought about the Shoreham fiasco without asking whether the NRC's safety findings were a mistake, or whether on the other hand Governor Cuomo, Suffolk County officials, and the public at large were wrong to claim that Shoreham would be an unacceptable risk. Yet it is impossible for a commentator to answer this question. Short of commissioning a comprehensive review of the NRC's safety findings by independent authorities with combined expertise in reactor design, construction, and operation, there is no way to second-guess the NRC's technical conclusions that the Shoreham reactor did meet federal standards and that those standards reasonably assure nuclear safety.

The safety record of American reactors, counting Three Mile Island as an educational near miss, indicates that the NRC was probably right about Shoreham, but even a perfect safety record (so far) leaves some room for reasonable doubt. Airliners sometimes fly thousands of hours with no accidents until an unlucky moment uncovers an unsuspected and fatal flaw. Closer to the point, the Soviet reactor program had a good safety record right up until the catastrophe at Chernobyl. American reactors do not have the dangerous design flaw that doomed Chernobyl, but it is not entirely impossible that they have some other serious vulnerability as yet undetected. The safety regulators assure us this is extremely unlikely, and as a practical matter we have to take their word for it. Regarding the safety of complex and potentially dangerous technologies, the public's only real choice is to rely on the official assurances of safety, as we do when we board an airplane or support the building of a nuclear plant. If trust is lacking, the system breaks down, no matter how technically sound the safety assurances happen to be.

The system broke down at Shoreham. What can and should be asked about that breakdown is whether the NRC went about licensing the plant in a manner that deserved public confidence. If the commission's actions gave the public legitimate reasons to doubt its fairness and objectivity, then the NRC should not have been surprised that the public refused to trust the agency's technical expertise.

This book takes the approach that distrust was the proximate cause, so to speak, of the wasteful outcome at Shoreham. There is nothing particularly deep about this conclusion as a general statement. Indeed, the commissioners and top executives of the NRC have over the years made many statements about the importance of public trust for the success of the nuclear power program, both preceding and following the Shoreham denouement. It can be argued, though, that the commission's actual tactics and policies disclosed a continuing agency distrust of the public, which in turn exacerbated public nonconfidence in the NRC. An important place to examine the existence and effects of this mutual distrust is in a major reactor-licensing proceeding, which is why a close study of the Shoreham story is worthwhile.

As this book will show, distrust at Shoreham did indeed go both ways. New York and Suffolk County distrusted the NRC, but the commission in turn doubted the competence and good faith of the state and the county in the emergency planning controversy. The NRC eventually expelled them from the proceeding. This action led finally to the long-sought licensing of Shoreham, but no good came of it in an atmosphere that was by then incurably poisoned with mutual recriminations.

Finding ways to increase public acceptability of nuclear power decisions is not just an academic exercise. Even though no one is now trying to get new nuclear plants licensed, in part because of the intense public opposition that could be expected, important decisions remain to be made on the renewal of existing licenses and on the disposal of radioactive waste. Ultimately, we must choose whether to abandon or hold on to the nuclear power option, weighing the disappointing economics and persistent safety concerns against the benefits of retaining an increasingly dependable way to generate electricity in baseload quantities with no air pollution or global-warming effects and no worry about availability of fuel. How the trade-offs balance out is not obvious. One hopes that these nuclear-related decisions will be approached with more realism than emotion, in a manner that is reasonable and fair both in actuality and in public perception. Those who have studied Shoreham will know that even the best-laid nuclear schemes can go awry, but they will also recognize mistakes that can be avoided.

**Licensed to Kill?**

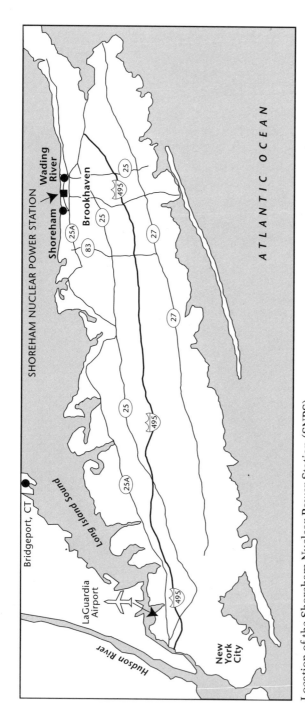

Location of the Shoreham Nuclear Power Station (SNPS)

# 1  A World-Class Fiasco on Long Island Sound

Although nuclear power provides a significant share of the nation's electricity (about 20 percent), enthusiastic predictions that by the end of the century America would have almost a thousand nuclear plants generating electricity "too cheap to meter" are obviously going to miss their mark. The commercial nuclear power industry is at a standstill. As all energy observers know, no new commercial nuclear power plants have been ordered in the United States since 1978 and more than 100 plants under construction in the 1970s and 1980s have been canceled or deferred. Although the United States now has about 110 commercial nuclear power reactors in operation, no utility is planning to build a reactor in the near future and some plants are prematurely shutting down.

Many reasons have been advanced to explain why the nuclear industry has faltered: high operating and construction costs; safety concerns; reduced growth in demand for electricity; design, construction, and operational problems; uncertainty about the management and disposal of radioactive waste; and an unwieldy regulatory process. Perhaps the most persistent problem is public lack of confidence in the management and regulation of commercial nuclear power. Despite the good safety record of the nuclear industry in the United States since the Three Mile Island accident in 1979, the polls remain roughly stable concerning public willingness to accept the risks of the nuclear technology. At the end of 1995, for example, a poll conducted by the Cambridge Energy Research Associates (CERA) indicated that about 50 percent of those polled opposed the building of more nuclear power plants anywhere in the United States. The numbers are little different from those shown in CERA polls conducted over the past five years and in Harris polls conducted in the mid to late 1980s.[1]

Despite the negative sentiment, many believe that nuclear power will make a comeback. Optimists in the nuclear industry foresee revived nuclear prospects due to a growing demand for electricity and increased concern about global warming and environmental protection. In particular, these forecasters envision a need for nuclear energy for baseload capacity because of emission limitations, limits on greenhouse gases, and an uncertain natural gas supply. The industry is planning to have a new generation of smaller, cheaper, safer units of standardized design reactors ready to go by the late 1990s. The industry has mounted a large public relations campaign portraying nuclear energy as a cleaner alternative to coal and other fossil fuels. And the new technology has received a note of affirmation from at least one environmental group that had in the past opposed licensing of nuclear plants.[2]

Nuclear power has received a boost from the government as well. The Bush administration favored the nuclear industry and extended financial and political support. President Clinton cut back on federal funding for research into nuclear plant design, but Hazel O'Leary, former secretary of the Department of Energy, expressed concern about keeping the nuclear option open. And Congress has passed H.R. 776, the Comprehensive National Energy Policy Act of 1992, which speeds up the nuclear licensing process. In accord with the new legislation, the NRC has adopted new regulations that simplify the licensing process for new plants and promote the use of preapproved, standardized plant designs.

But new legislation and new plants are not going to make adverse public attitudes disappear. As a National Academy of Science committee reported: "The public does not have a high degree of trust in either the governmental or industrial proponents of nuclear power."[3] This means, in effect, that the public has little faith that the nuclear industry or its regulatory body is willing to consider—and respect—interests and concerns that are different from their own, and little confidence that the decisions they make will be competently and objectively determined and substantively correct.

Distrust is not an uncommon complaint of government bodies in today's political environment.[4] But it is a particularly acute problem for an industry associated with what is widely regarded as a dangerous technology and for the regulatory body with jurisdiction over that industry. Nevertheless, neither the industry nor the government seems to have paid much attention to the issue of trust in commercial nuclear power. Many have paid lip service to the need for "public acceptance" to legitimate governmental action, but they have focused on ways to streamline and simplify the licensing process, to reduce the public's role in nuclear decision making, and to push through a program for waste disposal regardless of public opposition. These efforts, if not actually counterproductive, fail to address the basic problem.

To understand the consequences of public distrust for nuclear power, this book examines the experience of the Shoreham plant in Suffolk County, New York, the greatest nonaccident nuclear fiasco of the commercial nuclear power program. At the outset Shoreham was strongly supported both by the public and by state and local officials—by everyone, in fact, except for a small band of antinuclear activists—but the public mood changed over time, first to one of suspicion and subsequently to outright hostility. To observers familiar with the history of Shoreham, it seems clear that multiple mistakes were made in the course of building and licensing the plant. In fact, the mistakes were so extensive that the ensuing account might well serve as a primer on how not to build a nuclear plant. How blameworthy the mistakes were, who made them and why, and how they might be avoided in the future are less obvious matters, to be discussed hereafter.

Shoreham set many records during its lifetime, including a possible record for excess in the rhetoric used to describe it. The hyperbole is striking. Among other things, it has been called "every utility's nightmare"; "a symbol of the country's commitment to nuclear energy"; the "1980's catchword that describes all that's wrong with nuclear power"; and "a one word metaphor" for all of the reasons for the demise of nuclear power. James Watkins, secretary of the Department of Energy in the Bush administration, called Shoreham "the right energy source at the right time for the right place [whose] operation should stand as a symbol of the promise and potential of nuclear power in America." Peter Bradford, chairman of the New York Public Service Commission (PSC), called it "the ghost of nuclear power's failed past." And when the plant was abandoned, the *New York Times* termed Shoreham's dismantling "a blunder so monumental that, like the Pyramids, it may prompt future generations to marvel at the ruinous excesses of human folly."[5]

Shoreham differs from other troubled plants of the same era in many respects. It took longer to build, from start to completion, than any other nuclear plant. Had it operated, it would have been the most expensive commercial power plant in the nuclear industry in terms of dollars per kilowatt of capacity. It was a dramatic example of management incompetence and shoddy workmanship, having incurred the largest financial penalty—$1.4 billion—ever imposed by the New York PSC (or any other utility regulator in the United States) against a utility for defective construction and mismanagement. And it was the first full-sized plant to be decommissioned (or taken apart), earning the title "America's first stillborn reactor." Yet Shoreham could claim to be the safest nuclear power plant in the country, in terms of the risk of a nuclear accident, embodying more hardware changes and safety-related advantages than any other.[6]

Moreover, Shoreham represents an enormous waste, the largest investment in commercial nuclear power of time, money, and labor, all expended for naught. Its final cost was eighty times the estimated construction cost; its abandonment amounted to a $6 billion write-off for the American economy (roughly equivalent to each man, woman, and child in the United States throwing away $25). Additional amounts were spent by the Long Island Lighting Company (LILCO) on legal fees, consultants, and lobbying efforts to finance the fight to operate the plant and by Suffolk County and New York State, on the other side, to keep the plant from operating. Another billion or so was spent to finance the effort to dismantle it and dispose of the radioactive fuel. The ratepayers in the district to be served by Shoreham are bearing the major financial burden, including the cost of the plant's construction, the dismantling, and replacement electric power.

Shoreham is unusual in other ways as well. It provides the best illustration we have of the conflict between federal preemption and states' rights over commercial nuclear power. Prior to 1980, there was little question that the federal government—under the Atomic Energy Act of 1954 as amended—possessed exclusive jurisdiction over the operation of nuclear power plants. Although this is still the case in legal theory, Shoreham showed how a state could use its political and economic power to compel a private utility to refrain from operating a plant for which it had received a full power operating license from the federal government. For New York State, this was a spectacular, if Pyrrhic, victory. It marked the only time a state clearly won a fight to stop a contested nuclear plant.

Shoreham exemplifies too a flawed decision-making process for the licensing of nuclear power plants. The process is aimed at reaching a decision on a matter of great importance—whether a potentially dangerous power facility that represents a huge investment and that will affect more than a million people should be licensed to operate. Thus, there's a large stake in making a decision that is perceived as "correct" from all standpoints, including technical, economic, environmental, and policy considerations. As the succeeding pages will make clear, none of the contending parties—LILCO, the Nuclear Regulatory Commission (NRC), New York State, or Suffolk County—had confidence that the other participants were making an honest effort to achieve good public policy. Moreover, no party had confidence that the outcome would be "right" or that the process would lead to the correct decision. Consequently, each party felt vindicated in departing from the customary rules governing the process and engaging in questionable practices to achieve its ends.

Shoreham also illustrates the serious lack of any mechanism for reaching a compromise among opposing or contending groups. As the dispute intensified, each side boxed itself into an inflexible position in which only two major polar alternatives were considered: to abandon the plant even if this involved a sacrifice of a multibillion-dollar investment, or to operate the plant "as is" regardless of whether an effective response could be made in an emergency. No party to the controversy investigated the possibility of reaching a middle position that would address what had become the paramount question: Could a publicly acceptable state of emergency preparedness be achieved by actions that would cost less than the cost of scrapping the facility?

The give-and-take of an adversarial process was supposed in principle to protect the public interest, but at Shoreham it was not clear how the public was represented. The nuclear licensing process as structured by the Atomic Energy Act resolutely excludes a realistic approach to economic issues. The other interest groups in the Shoreham controversy seemed to think that, one way or another, they could pass the abandonment cost off onto somebody else. Lack of candor and hidden agendas on all sides compounded the problem. This was a poor way to make a complicated decision about an important, expensive operation.

Furthermore, Shoreham illustrates a divergence of views about how the public should be represented in the decision-making process. The federal government, the utility, the scientific community, and the professional associations, all believed that the decision about Shoreham's operation was primarily a technical one, to be based solely on the expertise and scientific judgment of the federal agency charged with legal responsibility to make the determination. These "experts" believed that they were the proper representatives of the "public interest," characterizing those who opposed Shoreham's operation as "adults who are making irrational decisions."[7] The state and local governments, the environmental groups, and the public believed that emergency planning (the major policy issue in the case) should be viewed as a legitimate subject for political discourse and compromise, to embody the views of those directly affected. A question for the future revolves around the role that might be played by groups outside the federal establishment. Should the role of the expert be strengthened vis-à-vis the public in an effort to minimize—and perhaps eliminate—obstructions and roadblocks in licensing nuclear power plants? Conversely, should the participation of affected citizens be strengthened, even if it slows the licensing process, in hopes of enhancing public confidence in nuclear power?

Whereas the peculiar characteristics of the Shoreham experience—the geography, the political environment, the public and private actors, and the contentious issues—created a situation that was unique in commercial nuclear power experience, there were enough elements in common with the licensing of other plants to stimulate a strong reaction from the industry, Congress, and the NRC. Major changes have been made to the licensing process governing the nuclear industry in the United States in order to standardize plant designs, streamline licensing, centralize decision making, and limit opportunities for public intervention. The industry has been given regulatory assurances that the licensing process would henceforth be "reformed and streamlined" to ensure that public intervention would not stop a completed plant from operating. Proponents believe that these will help to eliminate Shoreham-like challenges in the future and make the licensing process more predictable and certain.

This is doubtful. Shoreham shows that a licensing process, however streamlined, cannot by itself prevent waste of time and resources when the public lacks confidence in a proposed nuclear power plant. Obsessed with obtaining a desired outcome for the licensing decision, both LILCO and the NRC forfeited public confidence in the fight over Shoreham. Other nuclear plants have received operating licenses in the face of public distrust and outright hostility and gone on to generate electricity, if not the financial profits their builders anticipated. No doubt the NRC and LILCO expected that the fait accompli of license issuance would similarly resolve the Shoreham controversy to their satisfaction, if not the public's. The following chapters will explain why things did not work out that way at Shoreham. Shoreham shows that the most needed licensing reform is a serious commitment to building mutual confidence between the nuclear regulators and the public that has to live with the licensed plants.

The matter holds more than mere academic interest. The public controversy over nuclear power is a volcano that may be dormant but definitely is not extinct. Adverse reports at particular facilities and controversial general ideas (such as the proposal to use plutonium from nuclear weapons as reactor fuel) continue to arouse public interest and concern. There are no present proposals for new nuclear plants, but the operating licenses of existing plants will be coming up for renewal in increasing numbers in the years ahead. Many important decisions involving nuclear power remain to be made by the industry, the regulating agency, and the public. This book considers what can and should be done to keep distrust from corroding nuclear power decision making in the future.

# 2 LILCO's Early Enthusiasm and Errors

In the mid-1960s, LILCO's proposal to build a nuclear plant at Shoreham, Long Island, was welcomed by the state, the county, and the local community. But opposition quickly formed and then intensified when LILCO expanded its nuclear ambitions to Lloyd Harbor and later to Jamesport. A locally formed environmental organization vigorously opposed the Shoreham project at lengthy construction permit hearings. The opposition failed, but public distrust of LILCO emerged and would continue to grow during LILCO's effort to bring nuclear power to Long Island.

## A Power and Light Company Accustomed to Being Loved

For LILCO, geography has been destiny. The company provides power and light to an area that is long, narrow, and isolated. Long Island stretches 105 miles from Brooklyn and Queens at its rounded western end to lonely Montauk Point at the sharp eastern tip, but the greatest distance across the island from north to south is only 23 miles. The flow of people and commerce is generally channeled along the long east-west axis. Traffic tends to pile up as it approaches the congested boroughs of New York City, for there is no other way off Long Island by road or rail. Some Long Islanders will tell you that on a summer weekend afternoon when the expressways to the west are

Aerial view of the Shoreham Nuclear Power Station

jammed to capacity, there is (for motorists anyway) no way off the island at all.

Long Island residents, in common with the rest of the citizens of the New York metropolitan region, have opted for "a fragmented political system, rooted in local control, [that] at times seems stymied in dealing with island-wide or countywide issues."[1] Islandwide needs have to be met, though, if not by government then by private institutions, and among Long Islanders these have enjoyed varying degrees of respect and acceptance. The Long Island Railroad (LIRR) provides transportation, but if there was ever a time when riders did not ridicule the quality of LIRR service, no one can now remember it. There has been another islandwide institution, however, a better-respected one that tied together the otherwise fragmented Long Island community of small villages, towns, and all types of special districts, and this institution was LILCO.

LILCO generated and distributed low-cost power and appliances to over two million electric and gas customers throughout the two counties of Nassau and Suffolk, and to the Rockaway peninsula in the southeastern part of Queens, covering a geographic area of approximately 1,230 square miles. The time when LILCO found general favor in the community lies well within present memory. Many saw LILCO as not only a power and light company but also something of an eleemosynary organization, a "get-along-with-the-system kind of operation" that could be counted on to provide employment to job-seeking Republican Party members and to make contributions to worthy causes when needed.[2] Another observer noted that LILCO was "accustomed to playing a uniting role across the Island, a powerful civic role, a significant political role. It was accustomed to people respecting it for its fine civic functions. It had relatively low rates so it was accustomed to people thinking they were getting good value for their dollars. It was accustomed to being loved."[3]

On Long Island itself there is no coal or oil or other source of energy for producing large amounts of electricity. To make steam for the turbines that turned its generators LILCO burned fossil fuel, brought in large quantities from the mainland, unavoidably creating a significant amount of air pollution. Coal, the cheapest and most dependably available fossil fuel used for power generation, is also the dirtiest and the hardest to transport. Given the transportation problems unique to Long Island, LILCO had to rely primarily on oil, which was cleaner but more costly than coal and also increasingly vulnerable to the vagaries of the international situation.

The 1950s and 1960s were decades of enormous growth for Long Island. Between 1950 and 1960 the population more than doubled, from 950,000 to over 2 million. The economy was growing, and a sense of optimism abounded. LILCO's electricity sales grew as well, from 1.2 to 4.4 billion kilowatt hours. Between 1960 and 1970 the population increased by another 600,000 persons, making Long Island the fastest-growing segment of the New York metropolitan region. Electricity sales almost tripled, growing to 13 billion kilowatts. LILCO estimated that peak hour demand increased by more than 100 percent during this decade.

These were happy circumstances for a power and light company such as LILCO, but they were also problematic. The population of Long Island and the demand for electricity seemed to be on a rising curve that would never level off. More generating stations would be needed, more increasingly expensive oil purchased from undependable sources, and more air pollution tolerated. Was there any way out? Suppose Mephistopheles appeared from a cloud of smoke and offered LILCO's top managers a new kind of power plant, a nonpolluting plant that needed refueling only once a year with a low-volume, easy-to-transport fuel much cheaper (in the long run) than oil. In exchange for all this, LILCO would merely have to pay a little more at the start to build its facilities—and also, of course, sign over its corporate soul to a sophisticated new technology that only a few people (mostly outside the scientific establishment) were saying might be risky. Such a Faustian bargain might seem well worth considering and hard to turn down.[4] In real life LILCO confronted and accepted this bargain at Shoreham, failing to notice—or ignoring—the premonitory whiffs of brimstone.

### The Federal Nuclear Power Program and Government-Industry Promotionalism

Shoreham was a product of the first flush of enthusiasm for large nuclear plants, which began in the early 1960s when the Atomic Energy Commission (AEC) set out to convince electric utilities that nuclear power would be economically competitive with conventional generating facilities. The law governing construction and operation of nuclear power plants was a descendant of the Atomic Energy Act of 1946, which created the AEC and established a "program for Government control of the production, ownership, and use of fissionable material" (PL 585, 79th Cong. Section 1 (b)(4). In effect, the act created a federal monopoly over use of the energy confined in the atomic nucleus, energy that had literally burst upon the world's awareness less than a year before at Hiroshima and Nagasaki. In 1946 the idea that the controlled release of nuclear energy in reactors might someday be used to generate electricity was an attractive speculation but no more than that. The only practical use of the early reactors was to produce radioactive isotopes, most notably plutonium-239 for bombs. Under the Atomic Energy Act, private ownership and operation of such devices was deemed unlawful.

This highly restrictive regimen endured less than a decade. Concern about future shortages of conventional resources, together with fears that the United States might lose preeminence in nuclear technology if it did not actively promote civilian development of nuclear power, soon led to sweeping revisions in atomic energy law. The Atomic Energy Act of 1954 allowed and encouraged the construction and operation of nuclear reactors by persons in the private sector, pursuant to AEC licensing regulation. As an incentive to build plants, the AEC offered assistance with research and development costs and mitigated the cost of insurance covering liabilities stemming from nuclear accidents. Although plant owners had to pay for insurance coverage, the price of insurance underwritten by the government was far lower than the price of much less coverage issued by private insurance companies. In 1964, Congress amended Section 53 of the Atomic Energy Act to allow private industry to own special nuclear material (that is, plutonium and uranium enriched in fissile isotopes) used in nuclear fuel. Meanwhile, the government operated all enrichment facilities and developed research on waste disposal, particularly on high-level wastes.

The AEC's licensing policy was encouraging too, for it exempted nuclear reactors from prelicensing antitrust reviews. To stimulate interest further, General Electric (GE) and Westinghouse offered to build fully operational plants for a fixed price (known as turnkey plants). After contracting to build thirteen fixed-price plants (on which they lost money), GE and Westinghouse ceased offering guaranteed prices. Nevertheless, by the mid-1960s, there was great confidence in nuclear technology and a record number of orders for nuclear power plants. In 1966, twenty-four plants were ordered with a total capacity of 20 million kilowatts; in 1967, thirty units were ordered with a total capacity of 25 million kilowatts. The *New York Times* reported that the industry was "extremely bullish"; they expected 120–170 million kilowatts of installed capacity by 1980 (about 25 percent of the nation's total generating capacity). The AEC was optimistic also, estimating that 950 nuclear power plants would be operating in the United States by the year

2000.[5] Commentators have attributed this so-called bandwagon market, in part, to concern over environmental pollution from the conventional generation of electricity by fossil-fuel combustion.

New York State was confident about the future of nuclear power and led the way in promoting nuclear development by private industry and in institutionalizing atomic power–related agencies at the state level. New York was one of the first states in the mid-1950s to have a complete administrative and legislative program for the development of nuclear industries; it was the first to receive access to restricted data from the AEC to facilitate the commercial exploitation of nuclear material. New York adopted an atomic energy law in 1959 "to encourage atomic development and use within the state as fully as possible"; it established the Office of Atomic Development in 1959 and the Atomic Space and Development Authority in 1962 to implement the intent.[6]

Then Governor Nelson Rockefeller entered into a Memorandum of Understanding with the AEC in 1965 to avoid "dual regulation" against radiation hazards of activities licensed by either party within the state.[7] In 1967, Governor Rockefeller announced an $8 billion spending and expansion program by private utilities keyed to atomic energy that would double the production of New York's electric power by 1977, assure availability of "inexpensively produced and distributed electric power," and stimulate regional growth and development. New York's Public Service Commission (PSC) anticipated that nuclear power would be providing the major source of electricity within the next five years.[8] However, not everyone was enthusiastic about the anticipated nuclear future. The promotional posture by both federal and state governments during this early period led then representative Jonathan Bingham of New York to claim that the AEC and the state were promoting nuclear power without regard to safety considerations.[9]

## LILCO's Decision to Join the Nuclear Age Causes Opposition on Long Island

*Cheap, safe,* and *reliable* were the words LILCO used to describe nuclear power on April 21, 1965, when it announced its decision to build a five-hundred-megawatt nuclear plant in Suffolk County, which covers roughly the eastern third of Long Island. (Suffolk County is eighty miles long, and sixteen miles at its widest point, and is surrounded on three sides by water.) The proposal would ensure "large amounts of cheap electricity for Suffolk."[10] One year later, LILCO firmed up its plans by purchasing a 455-acre site overlooking Long Island Sound in the town of Brookhaven between Shoreham and Wading River, approximately fifty-five miles east of Times Square in New York City. Construction was to begin late in 1969, with service expected by 1973, and the cost would be about $65–75 million.

Frank Jones, who as deputy county executive for Suffolk County would become one of the leaders in the campaign by local government against Shoreham, recalls the general enthusiasm that greeted LILCO's proposal. "The first time the idea of nuclear power came forward, it was viewed like ice cream, apple pie."[11] Everyone jumped on board when LILCO announced its decision to build Shoreham. Wilfred Uhl, an electrical engineer and LILCO's president from 1978 to 1984, looking back at the event twenty-five years later, would note wryly: "We were not the only fools."[12] Public reaction in Suffolk County was generally favorable, and the political atmosphere was one of cooperation and support. County executive H. Lee Dennison spoke out in favor of the great new technology, saying: "I'm very happy about such a tremendous project coming into our area. . . . This is fine industry." Brookhaven supervisor Charles R. Dominy praised the forthcoming project as a "stupendous installation." The town's residents were delighted, too: the plant's contribution to town property taxes, based on current rates, was expected to exceed $1,500,000 annually. To the chief planner of Nassau and Suffolk Counties, "nuclear power was perceived as the least expensive, renewable, and clean."[13]

Encouraged by these reactions to the Shoreham announcement, LILCO saw itself surfing a wave of universal enthusiasm for nuclear power on Long Island and could not resist trying to stretch the ride. In October 1967, even before the Shoreham plans had been finalized, LILCO purchased property for a second nuclear plant site at Lloyd Harbor. Here, however, the nuclear future encountered a markedly different reception. This community was a beautiful peninsula, known as the Gold Coast of Suffolk County, with one of the highest per capita incomes in the county. It was home to many corporate chief executives, who were "horrified" to hear about a nuclear power plant on a prime waterfront tract in their community.[14] The move met with universal disapproval from the town's residents and precipitated Lloyd Harbor residents Ann and William Carl (a biologist and an engineer) to form the Lloyd Harbor Study Group (LHSG). LHSG began as a small conservation group of lawyers, scientists, and environmentalists opposed to nuclear power. It was a group destined to grow.

Ann Carl mobilized the opposition. She was a wealthy resident with a strong environmental sense and "anti-nuclear philosophically."[15] LILCO would abandon the Lloyd Harbor project in 1969, but by then the LHSG had become the vanguard of a movement against nuclear power anywhere on Long Island. Its membership of twenty-five hundred members subsequently raised a large sum of money (in the range of $250,000), brought together a large group of witnesses, and would spearhead the opposition to the Shoreham plant at the AEC's construction permit hearings. The LHSG determined that "as a matter of principle and credibility, we had to oppose the Shoreham nuclear reactor. . . . Once a nuclear power plant made its incursion anywhere on Long Island, there would be little chance of successfully opposing it anywhere else on Long Island." The LHSG's specific objections to Shoreham included "high population density, water pollution, earthquake susceptibility, and emergency evacuation difficulties."[16]

Thus LILCO's overreaching at Lloyd Harbor, right at the beginning of its engagement with nuclear power, needlessly stirred up troubles that would descend upon Shoreham later on. Ira L. Freilicher, formerly LILCO's vice president for public affairs (described as "LILCO's chief strategist"), called the purchase of property at Lloyd Harbor a "particularly stupid decision. . . . It was just a piece of land. There was never any hope of building there. . . . Nobody ever designed a plant there or filed an application. It couldn't even meet the criteria of the AEC at the time of purchase. It would not have been a good site for a fossil fuel plant."[17]

## A Disastrous Decision

The troubles LILCO would make for itself at Shoreham did not end with the abortive Lloyd Harbor adventure. For a while, the Shoreham project maintained the momentum acquired at its favorable start. On May 24, 1968, LILCO announced the filing of an application with the AEC for a construction permit to begin building the Shoreham nuclear power station. GE would design the boiling-water reactor system for the plant, and Stone and Webster Engineering Corporation would serve as the architect-engineer and would manage construction.

Early reports indicate that LILCO, in common with other utilities at this time, expected no greater difficulties with the construction of a nuclear facility than with a conventional fossil-fueled plant. There were a few cautious voices, however. Arthur Sugden, LILCO's vice president for engineering, wrote a memorandum in 1968 warning LILCO about the regulatory and financial uncertainties and the potential technical and labor problems attendant upon construction of a nuclear plant. He recommended that the Shoreham unit be deferred, pending a reassessment of the company's needs. Despite Sugden's warning, LILCO decided to move ahead.[18]

Simply moving ahead was not enough, however. In its enthusiasm for nuclear power, LILCO began to think "bigger will be better," and it made what would later be considered a crucial mistake. LILCO decided to enlarge the Shoreham plant from 540 to 820 megawatts, which would be four times more generating capacity than the biggest similar plant in operation. During the summer of 1968, demand for electric power on Long Island was up 20 percent over the preceding year, and the utility thought an increase in plant size would respond to new load forecasts projecting a continued increase in demand. The utility also believed that the larger capacity plant would be more economical than a comparable fossil-fuel plant since it would take advantage of economies of scale.[19] A former LILCO official noted that the enlargement turned out to be "a disastrous decision," but at the time it seemed based on "traditional utility economics. The basic cost was in the safety systems, and if GE now had an 820 megawatt, for not much more money that could give you more capacity, it made sense to proceed." Another LILCO official agreed, stating: "The company thought that a larger unit with the effective scale would have better cost associated with it. . . . I remember Sugden having me draw a curve of relative sizes and what you'd save from a scale factor on capital costs and using my curve to pick out a number to put in some kind of memos for some kind of documentation."[20]

The decision to increase the size of the plant caused a year's delay in the planning stage and necessitated a postponement in filing the

Shoreham application for a construction permit until May 1969. It also meant that commercial operations would not begin until May 1975 instead of 1973, the proposed starting date for the smaller reactor. It increased the estimated cost of the plant from $70 million to $217 million, the first of many such cost increases over the next fifteen years. The decision also led to design difficulties and delayed the start of hearings on the construction permit, a delay that happened to put LILCO directly in the path of newly arising requirements under the National Environmental Policy Act and the Federal Water Quality Improvement Act. The redesign increased LILCO's burden of responding to AEC regulatory changes.

As time went by, opposition to nuclear power was maturing. In retrospect, many knowledgeable persons associated with the Shoreham project believe it was a mistake for plans to stop and restart. Without the delay, Shoreham would have met less opposition and stayed out of environmental hot water. "It would have been easily licensed . . . run just fine. . . . Everyone would have loved it because it would have been there during the oil crisis."[21] But none of this was foreseeable at the time, of course. LILCO could reasonably fear that a decision *not* to enlarge the plant might later be termed a crucial mistake when Long Islanders found themselves paying more than was necessary for electrical power that was less than adequate. That LILCO's choice turned out to be condemned rather than praised was arguably just bad luck. And more bad luck lay ahead.

### The AEC-NRC Licensing Process and the First Shoreham War

After making the modifications required by the decision to expand Shoreham's generating capacity, LILCO submitted a revised construction permit application. In September 1970 an AEC hearing board convened to determine whether the application should be granted. Custody of LILCO's hopes for Shoreham now passed from the engineers and designers into the hands of the lawyers.

The Atomic Energy Act of 1954, as amended by 1970, provided for the issuance of a construction permit (CP) to a utility following safety reviews of the application by the AEC staff and a mandatory safety review by the Advisory Committee on Reactor Safeguards, an independent committee of scientists and engineers established by Congress in

1957 to advise the AEC on safety aspects of reactors. The act also required a mandatory public hearing on a CP before an Atomic Safety and Licensing Board (ASLB).[22] The licensing board consisted of three people (two members with technical qualifications as scientists or engineers and one with legal training and experience), who would review the AEC staff's recommendations regarding the CP application and make a decision—subject to further review by an Atomic Safety and Licensing Appeal Board and by the AEC itself—whether to grant the permit. Assuming that LILCO successfully negotiated this hearing, it could begin constructing Shoreham. Later, when the plant was nearing completion, AEC rules required LILCO to go through a similar process to obtain an operating license (OL), except that a public hearing would not be mandatory. At the OL stage the act required a hearing only if requested, either by persons who could reasonably claim they might be adversely affected by operation of the plant or at the initiative of the AEC itself.

Going into its CP hearing in late 1970 with an application already approved by the AEC staff, LILCO might reasonably have expected a fairly easy time. During the nuclear industry's formative years, the AEC was wont to regard the public hearings leading up to the issuance of a CP as somewhat of a formality, taking place after the important decisions had already been made by AEC regulatory staff. Because the staff and the applicant resolved their differences before the hearing was held, both appeared before the hearing board in support of the licensing application. Outside parties could challenge the application, either by participation as intervenors or by submitting a written statement of their position, but the public hearing process in effect at that time "rarely resulted in substantial controversy or debate of issues."[23] As one close observer recalled, the proceedings before the AEC were typically "little patty cake hearings in which truly helpless intervenors would have soothing words said to them, be allowed to be there for a couple of days and then go away."[24] In 1969 two AEC hearings on applications to build nuclear generating plants lasted only three or four days. In line with the prevailing pattern, the Shoreham hearings were expected to last about two weeks. Contrary to expectations, however, they developed into the longest and most contentious in AEC history, requiring seventy sessions, spread out over more than two years, to resolve the issues presented for ASLB review. These protracted CP hearings were appropriately characterized by one observer as the "first Shoreham war."[25]

The field generals in this war were the lawyers for the three principal parties—LILCO, the AEC staff, and the intervening LHSG. Of these, by far the most experienced veteran of the bar was the LHSG attorney, Irving Like, "a deceptively conservative-looking man with thick glasses and short grey hair, . . . who believed that nuclear power was one of the great societal ills—an article of religious faith."[26] With an undergraduate degree from City College of New York and a law degree from Columbia, Like had impeccable credentials. He was forty-six years old when the hearings began and had spent more than half his life practicing law. Martin Malsch, AEC counsel in the Shoreham CP hearings and until recently a deputy general counsel for the NRC, recalls that Like was "the best trial lawyer in the room throughout the proceedings."[27]

No carpetbagger in this Long Island controversy, Malsch also had excellent academic credentials from northeastern schools, including a law degree from the University of Connecticut, but he was eighteen years younger than Like and had little more than one year's experience as a lawyer. The Shoreham hearing was his first major appearance in a trial-type proceeding. W. Taylor Reveley III, representing LILCO, was a Princeton graduate with a recent law degree from the University of Richmond in Virginia. Malsch and Reveley could draw on the resources of a federal agency and a major utility company in making their case, but these two highly trained, relatively inexperienced lawyers would not find it easy to oppose LHSG and its shrewd, dedicated counsel.

The first of several reasons why the Shoreham CP hearing took much longer than expected was the innovative nature of LHSG's intervention. Like was opposing the plant on safety and environmental grounds, challenging the plant's design and its potential environmental effects as well as the AEC's licensing standards, but it appears he did not expect to prevail on the merits of his arguments. Like described his strategy at length in an article entitled "Multi-Media Confrontation— The Environmentalists' Strategy for a 'No-Win' Agency Proceeding." He saw himself arguing before a "kangaroo court," which was expect-

ed to issue an affirmative ruling to grant the CP. In such a no-win contest he advocated using the hearings "as an educational forum to alert
the public to the project's adverse effect on environmental quality . . .
and as a challenge to the technology itself." Appropriate issues to be
considered for comprehensive press and media coverage in such a forum included "the danger of accident which may discharge large quantities of radiation, the cumulative long-term effect of low levels of radiation routinely discharged, the problems of transportation and
storage of high level wastes, thermal pollution and other radiological
and non-radiological effects."[28] And Like did indeed raise these issues
over an extended period of time. Whatever effect these tactics might
have had on public education, there could be no doubt that they drew
out the Shoreham hearings and delayed the start of reactor construction.

All of the pro-Shoreham parties seemed unprepared for what tran-
spired at the hearings. To Dr. Vance Sailor, who led the Suffolk Scientists for Cleaner Power and Safer Environment (a group of about
eighty pronuclear scientists mostly from Brookhaven National Laboratory who intervened in favor of the issuance of a CP for the plant), it
appeared that "the NRC's rules at hearings were very loose and that the
system was an easy target for lawyers." He was "flabbergasted at the
testimony that was allowed by the ASLB." Sailor said:

[The LHSG] had an incredible number of witnesses. . . . They were testifying
about the effects of radiation and they couldn't define a rad or rem or explain
the difference between a curie and an absorbed dose. . . . I never want to go
through an experience like that again. . . . I was in a state of shock from a less
than satisfactory encounter with some of the world's leading authorities on
non-relevant subjects.[29]

Malsch agrees that Like's witnesses were unaccustomed to filing testimony at a formal regulatory proceeding:

They actually thought this was some kind of PR thing. Many of them had limited expertise in the field in which they were actually testifying. Practically
none of them had read the application or were familiar with Shoreham in any
detail whatsoever. They tended to be sort of general statements. I think the
LHSG made a decision that they didn't care what the board or the commission
said but they wanted to get maximum public effect out of their presentations.
They made a choice and I think it affected their credibility before the board.[30]

Exposed to a contested case for the first time, neither LILCO nor the AEC had been prepared for real opposition. A LILCO attorney observed that "the AEC was accustomed to relying on . . . expert bureaucrats operating in complete good faith, doing what they thought was in the country's best interest and basically taking a position with everybody else: 'Trust me. I'm expert. Don't make me explain. Don't unleash the media on me. Don't allow any politicians near me.'"[31] Malsch said the guidance to the AEC regulatory staff was then, as it is now, to "avoid undue delays. But exactly how you went about doing that was not very clear."

There were practically no contention rules, so that it was very difficult to define at any stage of the game exactly what the issues were. . . . Discovery turned out to be difficult because the issues were not well defined. . . . And the licensing board chairman had had no adjudicatory experience whatsoever in handling contested cases. . . . A very bad mistake, which impacted the whole proceeding. You could not get rulings from the presiding board requiring witnesses to answer cross-examination questions clearly. . . . This vastly complicated the proceeding and caused a level of frustration on everyone's part.[32]

Charles Pierce, president of LILCO during the hearings and a man described by *Newsday* as normally "quiet and self-contained," experienced containment failure as a result of the frustrating process. In the *New York Times* he charged that "fanatics and opportunists" in the name of ecology were blocking progress toward increasing energy production, including the construction of nuclear power plants.[33]

Malsch recalls that Like's strategy at the hearing, which closely followed the tactics he had outlined in his article, "was a major scandal at the time. It scandalized the agency and the Joint Committee on Atomic Energy in Congress, who thought the whole process was being grossly abused and manipulated but couldn't figure out what to do about it." Malsch was called down to the joint committee staff to explain why the Shoreham case was taking so long. As far as the joint committee was concerned, "[AEC] agency staff said the reactor was safe and what more is there? What was the problem here? Why couldn't they just issue a license? Why were these people concerned? The agency had spoken and that was the end of it."[34]

But it turned out that the agency had not spoken on everything relevant to the safety of the proposed plant, or at any rate had not spoken adequately. In May 1971, when the ASLB hearings on radiation, health, and safety at Shoreham had ended and "everyone was tired, especially the licensing board," the board had to reopen the licensing proceeding.[35] Generic questions had been raised about the adequacy of AEC standards for the design of the emergency core cooling system (ECCS), the engineering safety feature relied on to prevent a core meltdown if a valve or pipe failure led to a loss of coolant. The ECCS hearings lasted from November 1971 until January 1972.

### Enter the Environment

The delays entangling the Shoreham hearings seemed at times to be sent from capricious gods, bent on helping the Lloyd Harbor group to raise time-consuming issues. Such an intervention from on high was the *Calvert Cliffs* decision, handed down in mid-1971 by the U.S. Court of Appeals for the D.C. Circuit.[36] *Calvert Cliffs* had nothing directly to do with Shoreham, but it compelled the AEC to reexamine its approach to environmental issues, which the commission had previously been ducking. The consequences for the Shoreham CP hearing were profound.

*Calvert Cliffs* resolved against the AEC a lawsuit brought by a citizens group not unlike the LHSG, which opposed the Calvert Cliffs nuclear power plant then under construction near Maryland's Chesapeake Bay. The citizens group challenged the AEC's regulations for implementing the National Environmental Policy Act of 1969 (NEPA) in power reactor licensing proceedings. NEPA explicitly required federal agencies to address the environmental impacts of major actions significantly affecting the environment before making a final decision. Issuing a federal permit to construct a nuclear power reactor was clearly a decision that came under NEPA; and under AEC regulations the AEC staff had to consider environmental values in preparing its recommendation for action on a CP application. So far, so good. But the AEC's regulations also specified that the licensing board conducting the mandatory hearing on the staff's recommendation should *not* consider environmental issues unless these were raised affirmatively by parties to the proceeding.

In short, as the D.C. Circuit characterized the situation, the AEC's rule provided for the agency's NEPA responsibilities to be carried out entirely outside the hearing process. The regulatory staff's "detailed statement" on environmental issues would "accompany" the proposal to issue a CP through the agency review process, but there was no requirement that the initial decision maker, the licensing board, had to read it or even take it out of the envelope. Judge Skelly Wright declared succinctly that "the Commission's crabbed interpretation of NEPA makes a mockery of the Act" (449 F. 2d 1117). Wright ordered the AEC to draw up new regulations under which the commission would "take the initiative of considering environmental values at every distinctive and comprehensive stage of the process beyond the staff's evaluation and recommendation" (449 F. 2d 1119).

Applied to the Shoreham proceeding, *Calvert Cliffs* meant that the Shoreham licensing board had an affirmative duty under NEPA to make its own independent consideration of environmental impacts, whether or not Irving Like raised any environmental contentions. In reaching a decision, the board would have to balance the benefits of Shoreham against the adverse environmental impacts of building and operating the plant. To make this balancing possible, the AEC would have to make its own extended environmental review of the radiological and nonradiological impacts of the Shoreham plant, the first ever written for a nuclear plant. The AEC could not simply rely on certification from other federal and state agencies with jurisdiction over nonradiological matters (thermal effects on water quality, for example). To get the necessary information, the AEC required LILCO to file a far more comprehensive environmental report than it had previously done. All this, of course, took time. The second round of licensing board hearings—this time on NEPA issues—did not begin until December 1972, more than two years after the proceeding had originally convened.[37]

## Significance of the CP Proceeding

The hearings that constituted the first Shoreham war proved so unexpectedly long and burdensome that it is tempting, in hindsight, to see them as handwriting on the wall, a fatally unheeded warning to LILCO and the AEC that they should give up on Shoreham. But other proposed reactors would slog through long and acrimonious CP proceedings, be built and licensed to operate against intense opposition, and then go into more or less successful operation. Shoreham was to turn out different, but it is hard to make a persuasive case that the un- happy outcome years in the future could have been foreseen at the start.

One supposedly missed opportunity that Shoreham historians are wont to point to is that, among the many issues raised during the hear- ing, the intervenors contended that adequate provisions could not be made for evacuation in case of an accident. Ruling on this issue, the li- censing board stated that it had "no authority to pass upon the suffi- ciency of [evacuation] plans at this time or to deny the application on the grounds that they do not have a full blown plan for a facility that will not be in operation for several years." The licensing board saw no reason to pursue the matter further. The transcript also shows that Irv- ing Like (for the intervenors) inquired:

Was there any discussion of a public education program to notify the people of Long Island as to what would happen in the event of an accident at Shoreham and what each person must do to protect himself and his family?

Mr. WOFFORD [LILCO project supervisor]: It is our opinion that the indi- viduals don't have to do anything to protect themselves.

Mr. LIKE: Has LILCO made any study to determine the degree of par- paredness [sic] of the population of Long Island to cope with a major acci- dent at Shoreham?

Mr. WOFFORD: We made a review of this accident and what the implica- tions of it can be and we feel like there is no action required on the part of the public, that there is no great hazard involved and many, many of the charges that our opponents are making are absolutely false.

Mr. LIKE: You say you have made a review of the implications of an acci- dent?

Mr. WOFFORD: Yes.

Mr. LIKE: Is that contained in the Preliminary Safety Analysis Report?

Mr. WOFFORD: No, sir.

Mr. LIKE: Is it contained in any other written document?

Mr. WOFFORD: Not that I know of.[38]

Thus ended consideration of whether and how the public would be protected if a serious accident occurred at Shoreham. The emergency preparedness issue would prove crucial at the OL stage, rallying enor- mous public opposition to Shoreham, which probably settled the plant's fate. Obviously the issue did not get much of a hearing from the CP licensing board, but it should be noted that even if it had been thoroughly aired at the CP stage, it would have made no difference. Before the accident happened at Three Mile Island in 1979, neither the AEC nor its successor agency, the NRC, required much in the way of planning for serious accidents, which they regarded as incredible. The issue would have reemerged anyway at the operating license hearing, because the NRC's requirements for emergency planning were not fully established until 1980, following Three Mile Island.

Looking back on the Shoreham CP hearings, then, one might reasonably ask, What good came of them? Irving Like had gone into the proceeding asserting that no matter how well the LHSG made its case against Shoreham the CP would issue anyway, and of course it did. Like's goal, he had said, was to "alert the public" to nuclear dangers, but it appears that the LHSG witnesses showed at best a marginal knowledge of the subjects on which they were testifying. In fact, the argument was often heard—and not just at Shoreham—that the opponents of nuclear power were scientifically illiterate and could bring nothing valuable to AEC hearings. Citizens' groups such as LHSG that formed to contest the licensing of a particular nuclear plant obviously could not draw on the scientific resources available to the proponents of the reactor. Nevertheless, their participation in a contested hearing could serve a useful purpose, as the Shoreham CP hearing illustrated. One obvious benefit of the strongly contested hearings was to increase scrutiny of the plant itself by the press and by the NRC staff. Even Shoreham supporter Vance Sailor conceded that "the intervenors have forced an awful lot of people to go back and make a thorough reexamination of many factors we tend to feel very secure and confident about. Things like exposure levels to the population, safety factors that we have gotten to take for granted."[39] Martin Malsch would agree with this:

It is generally the case that when a case is highly contested, the staff, being human, tends to be a little more careful. And so I think all throughout the course of the Shoreham case, the staff, doing the safety evaluations, updating the Safety Evaluation Reports, were aware they'd be cross-examined, aware this was a contested case, and I think, as a result, a few more resources were put on the case and the staff tended to be a little more cautious. . . .

Somewhere during the course of the prehearing discussion, it's mentioned that there is this Nike missile base three or four miles off the site. I asked the AEC project manager what about this military reservation? Was it evaluated?

He answered there's nothing to worry about. Under the regulations, we don't have to do that sort of evaluation.

I said: Did you talk to the lawyers about that? Is anybody evaluating whether a missile could crash into the reactor?

He said: No. It's not within the scope of our safety mandate.

So he went back and the lawyers made them do the evaluation. They got a guy from the Department of the Army to do a big fancy probabilistic study of inadvertent launches. It turned out to be a contested issue. And it turned out that the probability of a problem according to the experts was so small that it was not a problem. The end result was that nothing changed about the plant, the site was still OK and the missile base was not a problem. But had there not been a contested case, I never would have known about it, we would not have asked the staff to do the evaluation, and no evaluation would have been done. Does that mean that safety was improved? In a sense it was because there was an evaluation. It could have been a problem. . . .

[Also,] there was a major controversy about the validity of our low-level radiation standards—the 500 millirem per year dose set for members of the public. And this hearing, along with a lot of other hearings on the same subject, eventually forced the AEC to revise its regulations and promulgate Appendix i, which kept releases to 5 millirem per year instead of 500 per reactor. It contributed to that overall effort to reduce in effect requirements on releases.[40]

## Licensed at Last

The seventy sessions of AEC hearings on the Shoreham CP, which included a hundred witnesses, continued intermittently from September 1970 until January 1973 and broke all existing records, making it one of the first major interventions by a citizen group in AEC history. The AEC issued a CP for Shoreham in April 1973, thirty-one months after public hearings had begun and fifty-nine months after the utility initially submitted its application in May 1968.

Irving Like and the LHSG had lost their battle, as they expected they would, but the "Shoreham war" was not necessarily lost. Opponents of Shoreham would get another chance to block the plant at the OL stage, if the resentment and hostility toward nuclear power that LHSG had aroused on Long Island during the controversial CP proceeding did not lose momentum and fade away. Perversely, LILCO proceeded on a course of action that assured this would not happen. Not content with its victory at Shoreham, LILCO promptly announced plans to construct two additional nuclear plants of 1,150 megawatts generating capacity each (a total three times the capacity of Shoreham) at Jamesport on the north fork of Long Island Sound, to lessen the utility's dependence on foreign oil. In embarking so quickly on this Jamesport adventure, LILCO seemed undeterred by the prospect of another round of prolonged and controversial hearings before federal and state bodies, a fresh spectacle that could have a major impact on LILCO's faded public image. Those hearings were still a few years in the future, though. With the Shoreham CP in hand, LILCO had a more immediate opportunity to show its fitness as a nuclear power licensee by doing a good job overseeing the construction of Shoreham. Unfortunately, it did not.

# 3 Constructing Shoreham and Making Enemies

A t Shoreham, construction proceeded slowly, with ever-escalating cost estimates and ever-receding completion dates. Mistakes, management problems, mounting expenses, and frequent rate increases contributed to public disaffection. By 1979, LILCO's blunders at Shoreham and overreaching at Jamesport had aroused so much antinuclear and anti-LILCO public opinion on Long Island that soon no one in elective office would risk defending the reactor. An enduring and effective opposition had been established that would eventually block the plant's operation. Lack of confidence in LILCO was a major factor in the Shoreham debacle.

## Primitive Kind of Managers

LILCO's first action under its newly received CP was to pour concrete for the reactor building. While waiting for the cement to set, LILCO officials were saying the plant would be operating by 1977 at a cost of $500 million.[1] Almost immediately, however, problems began to emerge that, it seemed, no one had thought about in advance.

New AEC regulatory requirements, adopted in 1971, involved a new American Society of Mechanical Engineers (ASME) piping code. Although piping was a significant construction activity, LILCO did not analyze whether the new regulations would increase the amount of piping to be fitted in the containment area. Nor did it consider designing a larger reactor building to house the new piping and restraints.[2] Consequently, the containment building became congested with piping and electric systems, which impeded subsequent construction work within the buildings. LILCO also failed to issue new piping specs and to procure the needed piping soon enough, causing piping material shortages and additional delays and costs.

Problems resulted from LILCO's failure to adhere promptly to the AEC's 1972 directive concerning protective measures to guard against damage caused by broken pipes outside reactor containments. Because of its delay in resolving this problem, LILCO did not receive the needed approval until 1977, four years after construction had begun. This too required redesign and replacement of a substantial amount of piping-related equipment.

As the pace of construction picked up, problems got worse instead of better. In 1974 the NRC, which had assumed AEC's regulatory functions, issued a regulatory guide establishing minimum distances for the separation of electric cables and conduit so that a single accident would not disable the electrical system and make a safe shutdown of the reactor impossible. After negotiating with NRC about the criteria it would use for spacing the cables, LILCO installed more than twenty-one thousand electric cables and conduits in violation of its own committed criteria. When the violations were discovered, LILCO continued to make installations, necessitating rework and reconstruction during the later stages of construction.[3]

Matthew Cordaro, LILCO's vice president for engineering during construction, blamed the problems in part on a late start. "After we got the CP, we realized we hadn't done as much engineering as we should have done in advance, and we were behind because we never knew what the licensing status of the plant was." Another LILCO vice president, Ira Freilicher, agreed that so much time was spent gaining the CP that the initial thinking on plant design was no longer consistent with NRC regulations. "In retrospect," Cordaro notes, "they should not have rushed ahead with construction but let the engineering catch up before they moved ahead."[4] The New York PSC later agreed with this assessment, pointing out that LILCO unnecessarily suspended engineering and procurement in August 1971 and failed to recognize that an engineering lead was needed to support construction once the project was restarted.[5]

LILCO insisted that the delays and management snafus were caused in large part by NRC regulations.[6] According to LILCO's Cordaro, "Changes in codes and standards took place daily, in large measure in response to a growing antinuclear sentiment and pressure on the AEC/NRC to take a more comprehensive look at nuclear power and improve on safety standards. . . . One of the reasons it [all] became as complicated as it did was because of the new standards, requirements, and quality-assurance procedures, which required much more complex systems for management of a facility like that . . . , and not knowing what tomorrow was going to bring, tearing out systems, putting in systems, quality-control inspection procedures, etc."[7] Wilfred Uhl, president of LILCO from 1978 to 1984, agrees that the changes were in effect costs imposed by the NRC, though not deliberately. Rather, they came about through the process by which nuclear plants were licensed and built:

You never had the assurance that the plans you filed are the plans you will be allowed to build. So throughout the project, which extended over ten years, the plant was continually changed because of changing NRC regulations. . . . It also destroyed the morale of the work force because on many occasions they had to take apart and demolish what they had already put together in order to conform with the new change.

The whole process of ever-changing regulations made it impossible for us to ever get a fixed price contract on anything from a contractor. Every other power plant I had ever been involved in building, and they're all oil-fired power plants, we would get the plans for the steam piping, we would hire a steam piping contractor, who would make a bid proposal on what it would cost to do the job, and we would select the lowest bid and have the job done. And the jobs always came in within the budget. This was the first time we ever walked out and said to a piping contractor we can't give you the plans because we don't know what the plans will be by the time you're going to build them. So, we're just going to ask you to supply the labor and the appropriate supervision for that labor and we will, in effect, pay you whatever it costs for you to build it. Now, we thought we selected good contractors and they were honest men. But when you put them in a position that this is a project where they'll be paid no matter what they do, and the other projects they're working on depend on how efficiently they do the job, I'm under no illusion that we got the best workers.[8]

Not surprisingly, NRC believes that the allegation of constantly changing regulations was "overblown. . . . LILCO had a lot of management and supervisory problems and there's no reason to believe Shoreham would have been treated any different from any other plant." Other utilities and construction firms found the NRC's requirements a nuisance but not "insupportable," and some were able to cope with them successfully.[9] NRC's former director of nuclear reactor regulation, Harold Denton, agrees: "For the utilities who do their jobs right, you never hear the charge of regulatory delay. For the utilities who do not do their jobs right, we're an easy target to blame." The New York PSC too found that LILCO was uncooperative and dilatory in responding to NRC questions and directives and found that they mismanaged several regulatory issues, which upped the project's costs.[10]

At the same time, several NRC managers of the Shoreham project pointed out that NRC paid more attention to Shoreham during construction than it did to other plants, because "there was a lot of litigation associated with the plant. If a reviewer had to prepare a defense against the issues, that's the reality of life here." Another NRC official told the New York PSC that the NRC required Shoreham to comply with the most stringent interpretations of regulatory guides and standards because "the NRC legal staff, based on the CP experience, wanted the strongest possible technical case for the Shoreham operating license hearing."[11]

Shoreham observers describe the GE Mark II reactor containment used at Shoreham as a "novel containment system" that was too small to support the engineering and construction needed for a reactor with a generating capacity of 820 rather than 540 megawatts.[12] Joel Blau, staff member of the New York State Consumer Protection Board (CPB), commented that entering the containment was like going "in[to] the middle of a bowl of spaghetti. Everything was so tightly packed. You just couldn't move around in there." Frank Murray, another New York official who visited the plant in 1983, said that "it didn't even look right. All sorts of things in different directions squeezed into place together as if in a shoe box."[13] And Wilfred Uhl admitted:

It turned out that there were many more feet of piping and electrical wiring that ended up having to go into Shoreham that hadn't been factored in. So then, in retrospect, the containment was too small; it was too crowded. The work force was hampered by that. . . . When the plant was finished, I never knew whether it was an operational plant, with all the safety systems that had been added. The more protective systems in place, the greater the chance that some minor malfunction will cause the plant to shut down.[14]

However, NRC did not think that "a crowded plant is by itself an unsafe one. You would just expect it to be down more often and have more maintenance problems."[15]

Most observers agree about the nature of the labor force—that it was unproductive and costly. An NRC attorney, asserts: "Any time you build on Long Island, you have labor problems. That's generic in that area." A LILCO vice president thought that "for most of its construction period, the Shoreham project was the only major one on Long Island, so the building trades were doing the best they could to keep the project alive, which meant 'slow the project.'"[16] Even more damaging, a PSC audit of LILCO's construction projects in 1978 found that craft workers at Shoreham (about twenty-five hundred in number at peak) spent an average of only 21 percent of their day in actual work (about 1.5 hours per seven-hour day), passing the balance of the day in starting late, taking long coffee breaks, waiting for materials, and leaving early. *Newsday* reported that the workers engaged in "sabotage, . . . work stoppages, . . . featherbedding, . . . and theft."[17] A former president of LILCO argues:

I'm not sure they goofed off more than any other project where you couldn't hold their feet to the fire. Talking to the construction superintendents right on the site, they would say: "Half the problem is that we can't give these guys a solid piece of work for them to do in the most efficient way." Yes, there was featherbedding. I think the situation was there to take advantage of and they took advantage of it.

Responding to allegations of Mafia-controlled operations, another LILCO official pointed out, "There was no particular Mafia influence unique to this job . . . ; just what was part of construction trades customarily—an attempt to milk the project for all it was worth."[18]

Observers agree too about the lack of management oversight and supervision. It became clear early on that LILCO could not exercise proper control over contractors and labor experiencing scheduling problems, equipment delays, cost overruns, and lagging worker productivity. The NRC project manager said, "Everything seemed to go wrong." Cordaro explains for LILCO: "We were on a learning curve. Stone and Webster had a lot of nuclear experience, a good reputation. We didn't want to criticize. . . . It was such a large thing that just grew and became so complex, that it was almost unmanageable. After a while, we realized that LILCO needed to tighten reins and control expenditures."[19] It was not until 1976 that LILCO replaced Stone and Webster with a management team, the Unified Construction Organization (UNICO), with LILCO in control and Stone and Webster as project engineers. However, NRC's director of nuclear reactor regulation visited the plant and was "singularly unimpressed with the involvement of the local management. I became convinced that the people in charge of the place really weren't up to speed."[20]

## Not Criminal Mismanagement but Malfeasance

Management deficiencies were clearly the problem that the PSC focused on in 1979 in deciding to establish a special proceeding to determine whether to disallow part of the cost of construction.[21] Management's inability to get cost and schedule under control was a major factor. Stone and Webster were operating on a "cost plus contract . . . with little incentive to meet target dates or hold down costs."[22] Instead of the cost figure of five hundred thousand dollars at the time the CP was issued, LILCO increased the cost estimate to 700 million in 1974, to nearly a billion in 1976, to one and a quarter billion in 1977, to a billion and a half in 1979, and to near two and a quarter billion dollars in 1980. LILCO extended the completion dates similarly on an annual basis; by 1980, commercial operation was estimated for 1983.

Each day's delay cost LILCO an increased amount in foregone investment returns. Wilfred Uhl stated, "The receding completion dates often contributed the biggest part to the costs because, in a project like this, the utility counts all of the interest on the money that's invested as part of the cost of the project." He added, "even to us, toward the end of the project, it no longer appeared that Shoreham would produce the economies that we had hoped for in the beginning." He pointed out that abandoning Shoreham was considered—and rejected—as an option in 1976, and again in 1979. "Assuming that we finally had our hands around the cost, it showed once again that the lowest cost alternative, from the point of view of consumer rates, was to continue with the project."[23] Another LILCO official agreed that LILCO's inability to control cost and schedule "gave the public the impression that the plant was sort of jerry-rigged," because when the schedule and cost estimates changed, the public impression was that no one knew what they were doing and this reflected on quality and quality assurance.[24]

Imprudence? mismanagement? or gross inefficiency? were the questions posed by the New York PSC in 1979 to determine whether the cost overruns of nearly 500 percent at the Shoreham facility were "prudently" incurred. Following the "most extensively litigated record in the PSC's history," the administrative law judges concluded that LILCO had "lost effective control over project costs and schedules" from the outset of construction in 1969 until 1984, when LILCO's chairman, president, and senior officials resigned. The PSC commissioners concurred, finding that "LILCO imprudently failed to establish, organize, and staff systems for controlling the cost and construction schedule at Shoreham, . . . mismanaged the engineering effort, . . . exercised inadequate control over design changes, . . . neglected to establish an effective system of measuring and reporting on construction progress, allowed unreasonably poor levels of productivity to persist throughout the project's history," and failed "to develop and administer an effective quality assurance program."[25] To penalize LILCO, the PSC ruled in December 1985 that about $1.4 billion could not be included in the rate base. This was the largest financial penalty ever imposed on a U.S. utility by a regulatory agency.[26]

## A "Much Too Ambitious Nuclear Program" Stirs Up the Opposition

During the years that construction problems were mounting at Shoreham, LILCO was not idle in creating trouble for itself elsewhere at Jamesport.[27] LILCO's corporate goal of constructing two 1,150-megawatt nuclear plants at Jamesport was bitterly opposed by antinuclear activists as unsafe, unnecessary, and overly expensive. In 1972, New York State had established a new agency to process applications for electrical generating facilities, the New York Board on Electric Energy Generation Siting and the Environment, which was responsible for approving all power plants within the state.[28] Hence, LILCO was required to obtain a Certificate of Environmental Compatibility and Public Need from the state, as well as a construction permit from the AEC. The New York board began to hold hearings in October 1974, and once more Irving Like was the principal attorney for the intervenors. This time he represented Suffolk County, which in 1977 adopted a strong stance against the Jamesport project on the grounds that the additional electrical capacity was no longer needed.

To LILCO, both sets of Jamesport hearings were "an exhausting, unnecessarily drawn-out and expensive example of progress being obstructed." To environmentalists and consumer advocates, however, the hearings were said to be "singularly successful . . . having provided probably the most complete examination ever conducted of the economic and environmental impact of a nuclear power plant." The hearings coincided with increased public debate on nuclear power nationally, with testimony on plutonium, reactor safety, radiation effects, sabotage, nuclear waste, evacuation of the reactor site, the merits of nuclear versus coal-fired plants, and the reliability of LILCO's projections.[29]

The underlying problem at Jamesport was public concern that LILCO was building too much nuclear capacity. For Irving Like, the central question was "whether these plants are needed." Even LILCO itself evidently had doubts on that score. In 1978 LILCO announced it would share the power generated by Jamesport with an upstate company, New York State Electric and Gas Corporation, a power importer. Like promptly charged that LILCO wished to become a "power exporter" and to "bail out NYS Electric and Gas." He also thought that LILCO was mistaken in pushing for the plants because "in the long run, if nuclear power forces the price of electricity up further and demand for power goes down, they may find themselves overextended and facing disaster."[30]

By this time, Long Island residents were indeed conserving energy, and electricity consumption had dropped. Peak electrical demand rose by only 2–3 percent per year from 1973 to 1978 instead of the 40 percent increase over the five-year period predicted by LILCO. At the same time, LILCO's estimated project costs for Jamesport were up, rising from 1.2 billion dollars in 1973, to 2.75 billion in 1977 and 4.5 billion in 1979, a fourfold increase. Even so, LILCO insisted the plants were needed, arguing that, without Jamesport, Long Island would face brownouts, blackouts, a limit on new electric customers by the mid-1980s, and economic disaster. As usual, the NRC licensing board agreed with LILCO, stating in 1978, "Jamesport is needed and the benefits far outweigh its monetary costs." The commission approved construction of the plants soon after. At the same time, the New York PSC found that Jamesport represented a "severe and unacceptable" burden on customers.[31]

Opposition to Shoreham escalated sharply during the Jamesport proceedings and spread to sectors of Suffolk County that had been relatively silent in the past. Farmers in the east end of the county became involved, because of their concern that the construction of high voltage transmission lines and towers in the potato fields would encourage the spread of golden nematodes, tiny worms that endanger Long Island's potato crops. Opponents included the governor, the Long Island farm bureau, the state department of agriculture, the state energy office, the state energy planning board, the Suffolk County executive, all of Suffolk's town supervisors, and the Suffolk County League of Women Voters. After six years of debate and about 150 days of public hearings, the New York siting board turned down LILCO's application in January 1980, because of "rising costs" and "regulatory uncertainty" caused by the Three Mile Island accident.[32]

Former LILCO officials now acknowledge the importance of Jamesport in sustaining a hostile public attitude to nuclear power and to LILCO. Ira Freilicher admitted:

LILCO held on to Jamesport much longer than it should have. . . . This was at a time when demand had fallen off and you could no longer justify the building of Jamesport on demand alone. You now had to justify it on grounds of oil displacement, various theories of fuel diversification, and based on predictions of what oil prices would become in a decade. And this was a difficult argument for the public and politicians to buy . . . even for us at LILCO.[33]

By this time, large segments of Suffolk County previously inclined to LILCO and Shoreham were "turned into a much more agnostic entity." Taylor Reveley described the prevailing sentiment:

A lot of people on Long Island were radicalized during the Jamesport struggle. Radicalized over a plant that was really not needed, much too expensive to ever be built, given the realities of LILCO's power situation. It made a fertile vineyard for the hard-core antinuclear activist. So if LILCO had just not tried Jamesport . . . from about 1974 to 1979 and fought so hard before the NRC and the NYS siting board to get it licensed, the public would not have been so aroused. It was a huge bitter struggle. . . . The war in Jamesport was basically waged with a lot of arguments about how fundamentally unsafe nuclear power is. . . . There are a lot of people with that sort of concern. And Jamesport brought all of that to the surface, kept it alive, kept it bubbling, kept it boiling during the interlude between the first and second Shoreham wars. It had exactly the same effect that reparations and depression had in destabilizing Europe between the First and Second World Wars. . . . I don't think the second Shoreham war would ever have been fought in the way it was had there not been the destabilizing, radicalizing Jamesport interlude.[34]

## No Other Utility Is a Beat for a Newspaper

*Newsday,* the local newspaper in Long Island, played an important role in shaping public opinion on Long Island and raising questions about LILCO's competence. *Newsday* had no New York City edition at the time. Its entire geographic area was Long Island "and the main thing that was islandwide was the LILCO bill [and] whatever plant LILCO happened to be building at the time."[35] Although the editorial policy toward Shoreham and nuclear power was positive, the plant became a principal target for the newspaper's investigative reporters. They were not described as particularly anti-LILCO; "it's just the inclination, particularly with a vocal opposition looking over their shoulders, for a headline writer to write the more sensational kind of thing, particularly the headlines."[36]

Thus *Newsday* gave wide coverage to the Shoreham CP hearings. LILCO spokesmen made best efforts to spend time with the press to ensure they understood the effects of terms such as "once-through cooling"—in which water is sucked into the intake canal from a body of water outside the plant (Long Island Sound, in the case of Shoreham), goes once through the plant, cools the condensers, and then is returned to the Sound as hot water. Their reward would be *Newsday* headlines such as "BILLIONS OF FISH EGGS KILLED."[37] Ira Freilicher noted: "So this heated things up considerably." There were times during the construction period that three to five articles about Shoreham appeared almost daily. As Karen Burstein put it: "Every time LILCO stumbled, it became more of a delicious target."[38] Taylor Reveley observed without rancor:

*Newsday*, in its commercial manifestations to sell newspapers, fed off LILCO and Shoreham and Jamesport for almost twenty years by being sensationalist and hostile. . . . It became politically suicidal to be pro-Shoreham after a while, and the main reason I think was that the media acted as an instrument of the opposition. . . . If you're talking about the big bad utility company, the big bad corporate entity, the only high technology that charges lots of money to people, has lots of lawyers, experts, against whatever it is, you're almost always going to get that kind of treatment.[39]

Negative publicity by *Newsday* culminated in 1981 in a lengthy series of articles titled "What Went Wrong?" by reporter Stuart Diamond, which highlighted among other things LILCO's scheduling delays, defective construction practices, and mounting cost and labor problems. A LILCO official said: "It telescoped fifteen years of mistakes . . . into a horror story. . . . And it had its most profound effect in that people thought there were a bunch of safety problems. . . . [It] gave [the] public the impression that Shoreham was a ticking bomb." It hurt LILCO in the public opinion polls and raised questions about its credibility. Another LILCO observer added: "the public saw the headlines, saw the pictures; if this gets so much attention, there must be something wrong."[40] Interestingly enough (as LILCO is wont to point out), a careful reading of the articles conveys the impression of an ill-conceived, overly expensive, and badly executed effort, but the articles did not question the plant's safety. At that time, it is fair to say, the plant was expected to operate.

## Public Enemy Number One

Shoreham became the ever-growing visible symbol of the institution everyone hated. No longer accustomed to being loved by the end of the 1970s, LILCO had become "public enemy number one" on Long Island.[41] When LILCO first began to build Shoreham, when most of the people in Long Island were not familiar with nuclear energy, LILCO told them that "Shoreham's operation will mean savings of hundreds of millions of dollars in future electric bills for LILCO customers." Then two things happened: "(1) Shoreham got out of control in terms of construction and costs and (2) the Arab oil embargo combined with

LILCO's freewheeling spending policies and people started to see annual double-digit rate increase requests."[42]

As the cost of the plant rose, so did the cost of the power the plant would produce. Some of the politicians saw the rising cost and thought it would no longer be popular to tie themselves to the Shoreham project. By 1979, recalls Wilfred Uhl, "we were already getting trouble from politicians who saw a cost problem coming up who then led a drumbeat of opposition based on fear of nuclear power."[43]

Although LILCO's rate increases because of Shoreham were small at the outset in terms of their impact on customers' bills, Long Island consumers failed to distinguish between Shoreham-related increases and those from a fourfold increase in fuel adjustment costs. Between 1970 and 1981, the New York PSC permitted LILCO to collect "tens of millions of dollars from customers for Shoreham before it operates."[44] By 1980, Shoreham's projected cost was $2.2 billion and its estimated operating date was 1983. *Newsday* reported that Shoreham was the most expensive nuclear reactor in history; while other larger reactors might have cost more in absolute dollars, the Shoreham plant was more expensive in terms of kilowatts of plant capacity, or $2,680 for each of its 820,000 kilowatts. Even the spokesman for the nuclear industry, Atomic Industrial Forum, reportedly stated that the average cost of plants expected to begin operating in the mid-1980s was estimated at only $1,200 per kilowatt of capacity. A spokesman for the New York PSC said: "What it all adds up to is this. Things go on bad just so long. Then they get worse."[45]

Indeed things did get worse. From 1973 to 1981, LILCO's financial condition steadily deteriorated. Less than 20 percent of LILCO's construction program was being financed by company-generated cash (the norm was reported to be 40 percent). LILCO's long-term debt doubled to almost $1.3 billion, and its annual interest payments rose from $34 million to $114 million by 1980. Standard and Poor lowered LILCO's bond ratings from AA in 1974 to BBB in 1980, and Salomon Brothers ranked LILCO number 94 out of 100 of the nation's largest utilities in terms of the quality of its earnings and cash flow. The value of a share of LILCO common stock, measured in constant dollars, fell nearly 50 percent between 1977 and 1981. LILCO's 1980 annual report showed that the company paid out more than twice as much in interest and dividends as it earned in cash. This meant that LILCO had to borrow just to meet expenses.[46]

By 1980, consumers in LILCO's service areas were outraged at paying huge rate increases to cover the massive cost overruns that would keep the utility financially liquid. Even Grumman Aircraft, the largest employer on Long Island (employing twenty-two thousand workers), became disenchanted with Shoreham because of the likelihood of a steep rise in electric rates. According to one observer, the battle cry was "If you kill Shoreham, you won't have to pay for it." In 1979 *Newsday* reported that 52 percent of Long Island was against operation of Shoreham, and 71 percent said they did not believe the threats of periodic blackouts.[47] A large coalition of antinuclear activists came together to form the Shoreham Opponents Coalition (SOC), which later testified against Shoreham in the OL hearings.

Anger was directed at LILCO as well for its "bunker mentality" and seeming insensitivity to public concerns. Faced with an unprecedentedly hostile environment, the utility failed to recognize the problems—technically, politically, economically—and did not appear to understand the need to relate to its customers. LILCO states that it responded to "what was customary for the industry. We looked to our neighbors, what other nuclear plants were doing, what was appearing to succeed in other areas, and used that as a yardstick."[48] Richard Kessel of the New York State CPB noted: "They were raising salaries left and right without caring too much about the people. They were hiding and misleading people about the completion of Shoreham."[49] In short, LILCO failed to adjust its behavior and activities to the changing political environment, in which classic demand for electric power was going down while the price of oil was rising; where it became more difficult to explain away the cost of nuclear power even if it was needed for oil displacement; where public opposition to nuclear power was mounting; and where Suffolk County was questioning its plans for expansion, all while LILCO's own financial condition was deteriorating.

And what did LILCO do in the face of mounting public disaffection? At Shoreham, as at Jamesport, the company resorted to threats—predictions of power shortages that escalated over time. LILCO warned of "periodic power blackouts and much higher fuel costs" by the early 1980s, and skyrocketing electric rates, "increasing hundreds of millions of dollars per year."[50] LILCO vindicated the choice of nuclear power at both Shoreham and Jamesport, assuming that "if they invested a lot of money, government would have to go along with them."[51] LILCO's "conviction that it could force the plants through" and its failure to respond to its critics played a large role in leading to negative public sentiment on Long Island. According to Bierwirth, its prevailing view was "You listen to us and we'll do the right thing for you."[52]

By the end of the 1970s, then, LILCO was in serious trouble. That this was not altogether its own fault was beside the point. Encouraged initially by both federal and state governments, and supported at the outset by the local community, LILCO was merely one of a large number of utilities that had been predisposed to believe nuclear power a panacea, and a relatively inexpensive one at that. Not alone in its choice to follow the nuclear path, however, LILCO nevertheless made more than its share of unwise business decisions: the abortive effort at Lloyd Harbor, stirring up a hornet's nest of environmental opponents, expanding the capacity of Shoreham without sufficient understanding of the problems, and persisting with the Jamesport project long after it was clearly not needed. Although the Shoreham construction problems might be considered "inadvertent" in that LILCO did not deliberately bring them on itself, they continued far too long. Poor management, faulty planning and supervision, rising costs, and inadequate attention to customer needs contributed to the general loss of confidence in LILCO. It is little wonder that the public lost faith.

# 4 Precursor of Disaster

## Three Mile Island

---

The accident at Three Mile Island in 1979 caused a sharp drop in credibility for the NRC and for nuclear power in general. It led to the adoption of new NRC regulations requiring emergency plans as a prerequisite for issuing an operating license. The emergency planning requirement became the central issue in contention at Shoreham and changed the focus of concern from a technical controversy to a political contest.

### Three Mile Island Takes the Bloom off the Nuclear Power Rose

On March 28, 1979, the most severe accident in U.S. nuclear power plant history occurred at Unit 2 of Metropolitan Edison's Three Mile Island nuclear plant (TMI) near Harrisburg, Pennsylvania. Technically, what happened at TMI was a loss-of-coolant accident (LOCA). The Babcock and Wilcox reactor at TMI, like nearly all U.S. commercial power reactors, was designed so that during operation the reactor core would at all times be covered by circulating water, which carries away the heat produced by nuclear fission.[1]

This heat is used to make steam, which turns the turbines that drive the generators of electricity. If something happens to cause a loss of coolant water (a break in a pipe is the classic example of a LOCA-triggering event), the reactor will shut down automatically—that is, the nuclear fission reaction stops. But if the reactor core becomes uncovered, and therefore uncooled, the heat from radioactive decay of materials in the core is enough to cause the dreaded "meltdown" of the nuclear fuel. A breach of the reactor-pressure vessel may or may not follow a meltdown, depending on how much fuel has melted and how fast the operators restore the level of cooling water inside the vessel. The possibility that a LOCA could actually get this far had previously been deemed remote. Reactors are designed with many backup systems to assure that, even if the highly unlikely pipe break or other leak should occur, emergency systems will inject more water into the reactor to make up for the loss and assure that the core remains covered.

At TMI a pressure-relief valve stuck in the open position, causing a moderately rapid loss of water from the system. This was a LOCA, though not of a classic variety. Nor was it a particularly threatening one. The emergency high-pressure pumps injected ample replacement water, exactly as they were supposed to. But then the reactor operators turned off the pumps. What no one had anticipated was that, in the particular sequence of events that happened at TMI, the operators, inadequately trained and misled by the reactor instrumentation, would mistakenly conclude that the system had too much water rather than not quite enough. With the emergency systems turned off and the relief valve still open, the core became partly uncovered and was severely damaged before anyone figured out what was going on.

Then, a couple of days after the onset of the accident, a bubble of hydrogen gas formed inside the reactor-pressure vessel, frightening almost everyone, including the NRC's scientifically trained chairman, with the possibility of a disastrous explosion. Babcock and Wilcox reactor engineers argued, however, that no oxygen could be present to form an explosive mixture, and their analysis proved correct. The hydrogen bubble diminished, and the reactor was brought to cold shutdown without further incident.[2]

Though TMI was an economic disaster for the utility, no damage occurred off-site, and no one was hurt. Radioactive releases to the environment were small and did not create a significant health hazard.[3] When it licensed TMI, the NRC had found "reasonable assurance" that the operation would not harm public health and safety, and in fact it did not. Nevertheless, the accident was an event that the commission's policy and regulations had treated as "not credible." What other, far worse nuclear plant accidents, also dismissed as not credible, might be waiting to happen next? TMI led to major changes in NRC requirements and practices and had a profound impact on the licensing of the Shoreham plant, particularly with respect to planning for emergencies.

### Emergency Planning Prior to 1979

Before 1970, the AEC had given little attention to emergency planning at nuclear plants or to the involvement of state and local governments in the emergency planning process. As more and larger nuclear plants were built and operated in the 1970s, regulations were issued that required each licensee to develop plans to assure protection of the public health and safety in the event of an accidental radioactive release.[4] Even then, however, a major off-site release of radioactivity associated with the most severe type of nuclear accident (so-called Class 9 accidents, involving disruption of the core) was regarded as so unlikely that it did not need to be covered by emergency plans.[5]

The AEC, and the NRC after it, relied primarily on siting requirements to protect the public, with small exclusion areas of about two or three miles surrounding each plant. The agencies required each OL applicant to include in its final safety analysis report (FSAR) arrangements for notifying state and local authorities of potential off-site releases, for assessing the off-site hazard, and for recommending protective measures should they become necessary. An OL applicant was also required to develop an emergency organization and emergency health services for its personnel within the boundaries of the reactor site. The commission did not require any planning for an emergency response outside the site boundary. The agreements between utilities and state and local governments were voluntary and were not imposed as a regulatory requirement in the licensing process. Hence, formal agreements were not always in place at the time licensing decisions were made, and the actual capabilities represented by the voluntary agreements were not tested. Looking back on those early planning provisions, Martin Malsch observed: "To the AEC, it was just a matter of making the right contacts and getting the right telephone numbers."[6]

During the late 1970s, serious questions were raised by concerned citizens, environmental groups, the General Accounting Office (GAO), and Congress regarding the actual capabilities of state and local governments to respond to radiological emergencies at commercial nuclear facilities. A report of the NRC/EPA Joint Task Force on Emergency Planning, published in December 1978, recommended that a deliberate effort be made to establish policy concerning the development of federal-state planning for emergency preparedness. A GAO study, issued coincidentally with the TMI accident, suggested that no new nuclear power plants should be permitted to operate unless offsite emergency plans had been concurred in by the NRC. A House of Representatives report entitled "Emergency Planning Around U.S. Nuclear Power Plants" urged the NRC to upgrade emergency plans and make them mandatory. The President's Commission on the Accident at Three Mile Island (the Kemeny Report) recommended, among other things, that approved state and local emergency plans be made a condition for licensing new power plants.[7]

### The Aftermath of TMI: More Power to the States

Barely five years old at the time of the Three Mile Island accident, the NRC came to Long Island with a blemished reputation. Although it had been created in 1974 by the Energy Reorganization Act as an independent, nonpromotional body, it was never able to shed the image that "it was the old AEC but with a different name."[8] The NRC's inheritance from its parent body included the AEC's regulatory personnel, its regulatory responsibilities, and its program, along with the five-member commission structure. NRC had the dual mission to perform the licensing of nuclear facilities and related regulatory functions and to provide adequate protection for public health and safety; but critics quickly charged the commission with a preoccupation with licensing schedules and with insufficient attention to overall safety issues. As former Representative Thomas Downey from Suffolk County saw it: "The licensing provision of the NRC is an advocacy process, not an objective one."[9]

The accident at the Three Mile Island nuclear plant in March 1979, "a traumatic, revolutionary event for the NRC," did away with the notion that a real nuclear disaster was such a remote possibility that it need not be planned for. The accident heightened public awareness of the potential consequences of nuclear power accidents and focused public attention on the need for plans and procedures "in case things go wrong in a way that you just couldn't imagine in advance."[10] To the NRC, the accident showed clearly that the protection afforded by proper plant siting and by engineered safety features had to be supplemented by assurance that the neighboring public could take protective actions during the course of an accident. Thus TMI encouraged the development and implementation of emergency response plans and preparedness measures. Reports by the various commissions appointed to investigate the accident led to the creation of the Federal Emergency Management Agency (FEMA) on the one hand and to the promulgation of new safety rules and regulations by the NRC on the other.

Another repercussion of TMI was a significant increase in the number of Long Island residents who opposed having a nuclear plant in their midst. Even formerly favorable or neutral observers started to look at Shoreham in a different light. About fifteen thousand people demonstrated outside the plant in June 1979. It was the largest demonstration ever held on Long Island. County executive John Klein traveled to Harrisburg to speak with Governor Thornburgh of Pennsylvania and came home troubled about the "appalling state of uncertainty and confusion" and discouraged about the possible development of a plan "based upon the geographic and population characteristics of Suffolk County and Long Island."[11] Wilfred Uhl, LILCO's president at the time of the TMI accident, pointed out how TMI changed the political climate:

Those who were opposed to the project, including those politicians who became opposed, always tried to paint LILCO as the devil. They became opposed to the rising cost. . . . And a lot of people who didn't believe those who were claiming that nuclear power was hazardous had their minds changed by the way TMI was handled. It created hysteria. It had an effect on the nuclear power business. That has turned out to be the most significant implication of TMI. Previously, we had pretty well persuaded people that even though this was costly, it really wasn't hazardous. From now on, people didn't believe us [LILCO] or the NRC that it wasn't hazardous.[12]

The principal issue affecting Shoreham in the aftermath of TMI was NRC's emergency planning rule and the peculiar way this rule was framed. Following the Kemeny Report, the rule required nuclear utilities to submit comprehensive emergency response plans to NRC, together with the coordinated emergency plans of state and local governments, as a condition for continued operation of a nuclear plant or in order to get an OL. Henceforth, no license could be issued without a finding by the NRC of reasonable assurance that adequate protective measures "can and will be taken in the event of a radiological emergency."[13] NRC's determinations would take into account the findings and determinations of FEMA regarding whether off-site emergency plans were adequate and capable of being implemented. Nevertheless, the final decision on the adequacy of both off-site and on-site emergency planning would be made by the NRC, not by FEMA. The standards that NRC and FEMA would use in assessing the adequacy of the utilities' plans were to be based on state-of-the-art knowledge in preplanning for technologically based accidents.

Although two congressional statutes—the NRC's 1980 Authorization Act, Section 109 (a)(2), and NRC's 1982–1983 Authorization Act, Section 5—provided NRC with the discretion to consider a plan drawn up by a utility if no local or state plan existed, the NRC rule contained no such provision. An assistant to former NRC commissioner James K. Asselstine pointed out that Congress was then motivated by the "Jerry Brown syndrome," or a concern that Governor Brown might withhold an emergency plan for one of the California plants and perhaps jeopardize its operation. NRC merely gave an applicant an opportunity to demonstrate, if regulatory standards could not be met, that deficiencies were "not significant," that "adequate interim compensating actions" were being taken, or "other compelling reasons" existed to permit plant operation.[14] "Supplementary information" to the emergency planning rule contemplated the participation and cooperation of state and local governments but acknowledged the possibility that some might be unwilling or unable to comply. If states were to decide against establishing the necessary emergency plans, they would forgo the use of nuclear power as a means of meeting their citizens' energy needs.

The NRC's determination of the health and safety significance of emergency plans and its conditioning of a license upon the existence and scope of the plans had the effect of involving state and local governments for the first time in the decision-making process relative to a safety issue in licensing a nuclear plant. NRC's former general counsel William Parler acknowledged this was a major policy change, pointing out that "prior to 1980, most of NRC's requirements had to do with questions that related to the adequacy of the nuclear steam supply system, the adequacy of a site, protection against earthquakes, tornadoes, etc., and the additional considerations that were introduced a decade before by NEPA. . . . These were issues exclusively within the province of the regulatory body." The emergency planning rule brought a new and potentially troublesome factor into the licensing process, as Parler noted: "If you provide an opportunity for a safety issue to be introduced, which involves political considerations as much as it does technical and engineering issues (as does off-site emergency planning), that introduces all sorts of complexities, particularly if the governmental bodies change their minds."[15] To put it less elegantly, the new emergency planning rule gave state and local governments a powerful new weapon: the governments could simply refuse to cooperate in the emergency planning process. Whether state and local government could thereby assert an actual power of veto over operation of an unwanted nuclear plant remained to be seen.

### Emergency Planning at Shoreham: A Critical Turning Point

The emergency planning requirement became the focus of controversy at a number of nuclear plants, but most decisively at Shoreham. Suffolk County and LILCO had been cooperating in the development of an emergency plan for Shoreham since the mid-1970s, resulting in a Memorandum of Understanding that was signed by LILCO and Suffolk County Executive John Klein in December 1979. Faced with an increased tide of negative public opinion and the defeat of several pro-LILCO and pro-Shoreham legislators, the new county executive, Peter F. Cohalan, who had approved the agreement initially, retracted his approval in 1981 and ordered the production of a revised Suffolk plan. The cost of the second plan ($245,000) was to be paid by LILCO.

LILCO paid $150,000 to the Suffolk County planning department in March 1982 as the first installment under the contract for developing the revised plan, with the balance to be paid upon the plan's completion. The plan was already close to 90 percent completed at this time.

Although the former Suffolk County planning director considered the March 1982 draft off-site plan (a four-inch-thick document of maps, charts, and directives) "unquestionably one of the best evacuation plans prepared in the United States,"[16] counsel to the county legislature called it "an apparent conflict of interest" and insisted that the down payment be returned to LILCO. The legislature then authorized the development of "a markedly different new radiological emergency response plan," which was to be submitted to federal agencies after it gained legislative approval.[17]

At about the same time, LILCO submitted the March 1982 draft off-site plan that had been prepared by the county (with LILCO's money) to the New York State Disaster Preparedness Commission (DPC) for review and possible state adoption. The county protested and filed suit for an injunction against DPC consideration. A settlement of this lawsuit resulted in a stipulation agreement between Suffolk County, New York State, and LILCO in December 1982, to hold DPC consideration of LILCO's plan in abeyance until February 23, 1983, by which time the county was to complete hearings on the revised plan.

The county's emergency planning task force, composed of "nationally recognized experts drawn from a range of pertinent disciplines," focused particular efforts on "the planning and preparedness problems caused by the special circumstances and conditions present on Long Island, such as the Island's elongated narrow shape, its severely limited roadway system, its quickly changing wind patterns, and its local demographic features."[18] Moreover, it was a nuclear emergency they were planning for, and this raised new questions. Frank Jones, the deputy county executive at the time, described in *Newsday* as "a big bearded man with a voice like a cannon and an appetite for combat to match," saw the problems:

Whether you could develop an evacuation plan for Long Island seemed a reasonable question. County had evacuation plans for hurricanes, all sorts of emergencies. . . . As citizen interest started to emerge, the county executive's office began to look more closely [at the plans] and some doubts began to arise. . . . They realized that the perception of a nuclear accident as an emergency was different and unique in people's minds. . . . Radiation is unknown, unseen, unfelt, . . . will make people react differently. . . . The more they got into it, the more complicated it became. . . .

A traffic engineer doing work on emergency planning in Suffolk County suggested, "in an emergency, make all roads one way out." . . . But what happens if kids are at school? Answer was buses would go and bring kids home. [However,] . . . in Suffolk, 65 percent of the school buses are driven by mothers whose kids go to school. What would they do in case of a nuclear accident? What would teachers as mothers do? [Also,] . . . if a husband is at work and he has the car, wouldn't he have to come back in to get the kids out? So the question arose: "How do you control parents wanting to rescue their own kids?" And the answer was: "Maybe use deadly force to keep order." A representative of the police department was doubtful. "What makes you think the police will be there? We'll be home taking care of our family."[19]

## You Can't Evacuate Long Island

The principal dispute between the experts retained by the county and those working for LILCO related to the size of the area for which planning should occur in the event of an accident. A significant section of NRC's emergency planning rule dealt with requirements for "emergency planning zones" (EPZs) around each nuclear plant: a "plume exposure pathway EPZ" of about ten miles in radius and an "ingestion pathway EPZ" about fifty miles in radius for dealing with food and water that might become contaminated. (The "plume" refers to the cloud of radioactive gas and suspended particles released and blown downwind from a disabled reactor. An accident that released a significant plume would have to involve not only melting of the nuclear fuel, which occurred at Three Mile Island, but also a breach of the containment structure, which did not. The NRC continues to believe that the odds against such an event happening are immensely high at any nuclear plant built and operated in compliance with the NRC's safety regulations.)[20] The rule provided flexibility, stating that "the exact size and shape of each EPZ will be determined by emergency planning officials after they consider the specific conditions at each site."[21] This provision surfaced prominently later on as a source of contention at the NRC Shoreham hearings.

The NRC requires an emergency response plan that includes "a range of protective actions . . . for emergency workers and the public" inside the plume exposure EPZ. Despite this "range" language, for a person in the anticipated pathway of a radioactive plume there are really only two basic protective actions: either take cover ("sheltering") or get out ("evacuation"). Obviously, getting out provides the surest protection but only if one is assured that evacuation can be accomplished before the plume arrives. The worst situation would be an evacuation that started too late or one that proceeded too slowly (because of massive traffic jams, for example), so that persons attempting to escape are overtaken by the radioactive plume with nothing for shelter except stalled or overheated cars and buses.

In line with NRC requirements, the LILCO plan envisioned a fairly simple and straightforward emergency plan that included evacuation from an EPZ ten miles in radius. The plan estimated that evacuation would be an "orderly process" and, depending on wind direction and the seriousness of the accident, would be completed in about six hours.[22] The Suffolk plan, by contrast, was detailed, elaborate, and pessimistic. Based on probabilistic risk assessments, extensive analysis of the types of accidents possible at the Shoreham plant, and the consequences of such accidents, it called for an EPZ for airborne exposures twice as large as that called for by NRC regulations.

Suffolk County's proposed doubling of the size of the EPZ had its basis in the peculiar geography of Long Island. The EPZ of the NRC's regulations was a generic concept that presupposed a relatively symmetrical geography—"an area about 10 miles in radius" (10 CFR 50.47 [c][2]). Evacuation would always proceed along a direction radiating outward from the facility. As evacuees moved farther from the plant, more space would open up and movement would become progressively easier. Any emergency response that might be needed at distances of ten miles or more from the facility could be accomplished on an ad hoc basis without a need for advanced planning.

But Shoreham did not fit this convenient picture. Half of any circular EPZ centered at the plant would be out in Long Island Sound. A Dunkirk-like evacuation (with a flotation of boats) across the sound to southern New England was not a reasonable option, and the circular symmetry model failed at the outset. Because of Long Island's size and shape, its seasonal increase in population, and its congested road system, the county concluded that evacuation planning could not be limited to areas within ten miles of Shoreham. In the county's view, a planning zone of less than twenty miles radius around the Shoreham plant for airborne exposures would be inadequate to protect the health, safety, and welfare of its residents. The Suffolk EPZ would take between fourteen and thirty hours to evacuate, depending upon weather and the size of the release.[23]

At legislative hearings on the Suffolk plan, one county commissioner after another testified against the feasibility of evacuating this area quickly in an emergency situation. An evacuation, they said, would result in a "major traffic disaster," in which more than a million people would try to move.[24] Roads would be blocked, drivers would panic, cars would run out of gas and stall, children would scream, and police officers and firefighters would leave their posts to assure the safety of their families. To complicate the evacuation, a county-sponsored attitude survey of Long Island residents indicated that there would be a "shadow evacuation" phenomenon, in which "tens of thousands of people outside the evacuation zone would spontaneously take to the roads, blocking those in the areas actually threatened by radiation."[25]

This shadow phenomenon was another consequence of Long Island's geography, because, for anyone fleeing a nuclear accident out on the island, all roads to safety led west. For islanders east of Shoreham, evacuation could mean initially going *closer* to the plant, and entering the EPZ if necessary, on their way to refuge in the western part of Long Island or the mainland. The county's view of the shadow phenomenon was buttressed by sociological studies and by a Suffolk poll disclosing that, based on the severity of the accident and the residents' distance from the plant, 25–50 percent of all residents of both Suffolk and Nassau Counties would try to leave the area.[26] The findings leaned on the TMI experience, in which—following official advice to evacuate young children and pregnant women within five miles of the plant—about 60 percent of all people within five miles, 40 percent of those within five to ten miles, and 32 percent of those in a ten-to-fifteen-mile radius left voluntarily. Commenting on this situation, Governor Thornburgh reported: "When you talk about evacuating people within a 5-mile radius of the site of a nuclear reactor, you must recognize that that will have 10-mile consequences, 20-mile consequences, 100-mile consequences."[27]

The county commissioned sociological studies that polled about twenty-five hundred Suffolk residents to investigate their intended behavior in the event of an accident at Shoreham. A survey of school bus drivers, ambulance drivers, and volunteer firemen found that a majority of those questioned would experience a "role conflict," that is, they would assure the safety of their own families first before performing their customary evacuation duties.[28] Given these findings, the county contended that large numbers of emergency workers would not be available for an evacuation.

The county concluded that evacuation was not acceptable as a protective action. And "sheltering," the alternative option, would expose Suffolk residents within the EPZ to unacceptably high doses of radiation. "Assuming a worst-case wind toward New York City," the deputy county executive said: "accident induced latent cancer fatalities within the 20-mile zone range from none at all, given sufficiently rapid evacuation, to 1,800 fatalities. . . . Beyond the 20-mile limit, [there would be] 7,000 latent cancer deaths."[29]

Henceforth, "You Can't Evacuate Long Island" became a popular rallying slogan for Suffolk residents. Although LILCO made detailed traffic estimates and reported that the area for plume exposures could be evacuated within reasonable time intervals, Suffolk residents did not believe it. The residents knew they lived in the most congested suburban area in the country and that, on a summer day, there was no way to move. They knew that there was ample justification for calling the Long Island Expressway, one of the principal east-west routes along which evacuation would be conducted, "the world's longest parking lot." Suffolk County residents felt vindicated some years later when Donald Regan, then White House chief of staff, stated publicly on "Meet the Press": "As one who lived in Long Island for quite a few years, [I know] there's no way to get off Long Island on a Sunday afternoon."[30]

The Suffolk County executive submitted the plan to the county legislature in December 1982. Following public hearings that presented a "parade of horrors . . . of thousands of cancer incidences and fatalities," the county legislature agreed with the executive that the risks were "unacceptable" and said it would be "impossible to prepare a local radiological emergency plan that would protect the public in the event of a serious nuclear accident at the Shoreham plant."[31] The legislature voted to terminate all emergency planning efforts and requested that the NRC forgo any further licensing action for Shoreham. In a prescient op-ed piece in the *New York Times* titled "Don't Operate Shoreham," the county executive supported this request, pointing out the two choices then confronting Shoreham's future: "continuing confrontation and interminable litigation or abandonment of the plant and a negotiated economic settlement."[32]

## Emergency Planning: A Political Ploy?

Not everyone thought the emergency planning issue was a genuine concern. Shoreham supporters believed then—and still do—that the concerns raised by the county were a "phony issue," a "sham," a "pretense," providing a convenient rationale for those who wished to block the operation of the plant for reasons having more to do with politics than with public safety.[33] They maintained that "emergency planning was contrived, politically manipulated, by the Suffolk legislature and executive," who saw it as a "useful issue." (LILCO claimed that Suffolk's opposition to Shoreham was a diversionary tactic, calculated to draw attention away from a well-publicized scandal concerning the financing problems of the Southwest Sewer District.)[34]

Whatever their motivation, opponents of Shoreham had found their most effective issue in emergency planning. Matthew Cordaro, LILCO's former vice president for engineering, pointed out:

Irving [Like] was the first guy to see the Achilles heel in emergency plans at a very early stage. He had a feel for the fact that emergency planning would have public appeal. . . . And I always knew that once they made a big thing of the fact that there was a need for . . . the most elaborate emergency plans that existed for any industrial or technological type of operation, then you really fuel public concern. Once you admit the need to do that then people think "this thing must be dangerous."

Another LILCO vice president agreed:

Emergency planning is such a good public relations issue because a nuclear plant owner can't win an argument on emergency planning because, in the public's mind, you're talking about *emergency, emergency, emergency,* and all they can think of is evacuation. So every time we won an argument, we lost. And it was an issue that all we could do was to stay away from it, from a public point of view. But of course, we couldn't totally stay away from it.[35]

Cordaro conceded, however, that "Irving [Like] really believed in this. He was a Long Island boy. This was a cause." Cause or not, in

1980 Like was dismissed as Suffolk County's special counsel on energy-related matters by county executive Peter Cohalan because he had backed the reelection campaign of the opposing candidate.[36]

Like's successor, Herbert Brown, hired by the Suffolk County legislature in February 1982 to represent the county at NRC proceedings, was not a Long Island boy. Sometimes described as a carpetbagger, Brown had established his reputation as an antinuclear lawyer on the other side of the continent, where he represented the state of California in opposing the licensing of the Diablo Canyon plant. He was said to consider Shoreham as the Diablo Canyon of the East. "Just as Diablo was put in the wrong place two and one half miles from a major earthquake fault, the Shoreham plant probably should never have been built where it was, because it is in an area that may be impossible to evacuate."[37] Brown had experience tangling with both the AEC and the NRC over emergency planning. He was reported to regard it as a winning issue with the electorate. "He knew how to stroke the right chords and to play the game with the press, make the right statements at the right time, and to delay things and make things complex." Having the lawyer's instinct for the jugular, Brown "knew that the magic formula for killing Shoreham was two words, 'emergency planning.'"[38]

Peter F. Cohalan, who had defeated fellow Republican John V. N. Klein for the position of Suffolk County executive, was especially vulnerable to the charge of opportunism in the Shoreham controversy. Long known as a Shoreham supporter who did not wish to do anything to delay Shoreham's operation, Cohalan was reported to have made "a back door entry to the forefront of the anti-nuclear movement." Suffolk's planning director said that the "clincher" came for Cohalan when he went to visit the plant and LILCO refused to let him in. It was at this point that Cohalan is reported to have said: "These son-of-a-guns must be hiding something."[39]

Deputy county executive Frank Jones warned LILCO that "there is no political benefit for Peter [Cohalan] in making a settlement with LILCO. . . . From now on, Peter is going to be on the left of Jerry Brown." However, nobody at LILCO took it seriously because "Cohalan was thought of as a conservative Republican and he'd never burn his bridges with the business establishment on Long Island." Others interpreted his opposition differently: "[He] found an issue for his next campaign [and saw it] as a means of getting a foot on the ladder for governorship."[40] Jones disputes the notion that he and Cohalan were motivated by political concerns, stating: "It was a difficult political decision to come to. We are Republicans. We believe in nuclear power. It was a case, not a cause, a case of a nuclear plant being in the wrong place."[41] Bernard Bordenick, who had succeeded Martin Malsch as the chief NRC attorney on the Shoreham case, would not have agreed. He thought, as did almost everyone at NRC:

The politicians knew which way the wind was blowing. They sensed the majority of the people didn't want it. . . . They threw up all these bogeymen. . . . Volunteers are going to stop doing their job. . . . If the emergency plan for Shoreham, putting aside the geographical configuration, isn't any good, none of the emergency plans in the United States are any good. Because they all rely on volunteers. I go with the sociologists who say, "People, when the disaster strikes, rise to the occasion."[42]

But one could not "put aside the geographical configuration." And some high-ranking politicians saw real problems in the evacuation plans.

## We Are Not Going to Impose Our Will on Local Government

Mario Cuomo became governor of New York at the beginning of 1983. He had been in office two months when he received a call from Dr. David Axelrod, New York State health commissioner, telling him that the court-appointed stay for the submission of the Shoreham emergency plan was due to expire the next day. Did he want the plan submitted to Washington? The governor responded that the state Disaster Preparedness Commission would not review the LILCO plan or approve any evacuation plan that Suffolk County did not support. According to state official Richard Kessel, then head of New York's Consumer Protection Board (CPB), Cuomo was a "strong believer that this was a decision made by Suffolk County and it wasn't up to him to order the county to do something different. The best official to make the decision about whether or not you could evacuate an area is the person who's in charge of that area. Even if he felt the other way (and he didn't), he would not have imposed his will on Suffolk County." Frank Murray, then deputy secretary to the governor and his advisor on energy issues, agreed that "it reflected the governor's respect for local government. . . . Not at all surprising that a man who had spent so much of his career in the private sector representing the little guy against government—be it city, county, or state—on an issue like this would defer to the sensibilities of local government."[43]

Cuomo detractors questioned then—and still do—the seeming inconsistency of the governor's supporting nuclear plants in upstate New York and the two Indian Point plants in Westchester County near densely populated New York City at the same time as he opposed the opening of the Shoreham plant. They say: "You can't trust nuclear power plants at Indian Point and not in Suffolk County."[44] They point out that in May 1983 the governor used New York State resources and personnel to compensate for Rockland County's failure to approve an emergency plan for the Indian Point plants; if he had been willing to do for Shoreham what he did for Indian Point, the issue would have been resolved.

The two situations, however, were not completely analogous. Rockland County was the only one of four counties in the Indian Point EPZ not supporting emergency planning, and Rockland expected to develop a plan in the near future. (The remaining three counties had developed emergency plans.) Furthermore, Cuomo supporters point out that at Indian Point, "you wouldn't have to evacuate anybody directly to the plant; . . . you could evacuate in all kinds of different directions and be safe. In Long Island, the only way to evacuate the northern tip, short of building a bridge to Connecticut, is to bring them back in the direction of Shoreham unless you're prepared to write them off." There are practical considerations as well. A former NRC commissioner suggested that "it is harder to take a plant or two plants out of service than it is to oppose one that has not come in yet just from a pure power-supply-planning standpoint."[45]

Indian Point was not the only reason for questioning the governor's sincerity, however. Cuomo hedged on state action toward Shoreham for months after he took office. At the same time that he ordered the state to withhold approval of LILCO's emergency plan, he expressed his willingness to mediate an accommodation between the county and LILCO.[46] A few months later, the governor appointed a fact-finding commission to study the problem, raising the possibility that he might reverse his earlier stance. He supported a strong federal role in emergency planning, stating that "the federal government must assume responsibility for the development and implementation of off-site preparedness training at nuclear power plants throughout the nation," including direct provision of funding and personnel.[47] Soon after, Cuomo asserted the supremacy of state government, saying that the state should have the final say over emergency plans. If the state says no, the federal government should shut the plant down.[48] Subsequently, he opened the door for Shoreham's operation, stating that New York might participate in emergency planning around Shoreham if the federal government were to meet his demands. The *New York Times* termed Cuomo's contribution a "sostenuto of equivocation," stating that the ultimate issue at Shoreham was not "safety" but "leadership."[49]

## An Impossible Mission

It is little wonder that the governor appeared to vacillate. Faced with divergent views among his top advisors, many of whom believed that the Shoreham plant was too close to completion to stop, the governor adopted a time-honored method for putting off decisions: he appointed a fact-finding commission. The commission consisted of a broadly representative group, "with expertise in public health, consumer affairs, nuclear technology, business affairs, suburban studies, economics and regulation," whose mission was to "develop clearly derived, reliable and objective information on the economic costs and safety" of the facility.[50] Dr. John H. Marburger, a physicist and also president of the State University of New York at Stonybrook, served as chairman.

Marburger's panel was divided among opponents and proponents of the Shoreham plant. Of the twelve members, three were avowed critics of Shoreham: Leon Campo, executive assistant for finance of East Meadow, Long Island, public schools, and chair of the Peoples Action Coalition; Marge Harrison, co-chair, Long Island Progressive Coalition; and David J. Willmott, editor and publisher of *Suffolk Life*. Three members—Alfred E. Kahn, economist, of AEK Associates; Dr. Herbert J. Kouts, chairman, Department of Nuclear Energy at Brookhaven; and William J. Ronan, manufacturing executive and former chairman of the Metropolitan Transportation Authority—tended to believe that Shoreham could be operated safely and that the economics favored operation rather than abandonment. The remaining members included Dr. David Axelrod, NYS health commissioner; Karen Burstein, president of the NYS Civil Service Commission; Dr. Paul Marks, president of Memorial Sloan-Kettering Cancer Center; and Hugh Wilson, a professor at Adelphi University. Both the NRC and the FEMA supplied nonvoting representatives to serve as panel members.

If the governor sought strong recommendations to shape his views about potential state action, he must have been disappointed when he received the Marburger panel's report. There was little direct guidance in the report either to support Cuomo's early decision to support Suffolk County's position in the Shoreham controversy or to give him cover if he wanted to change his mind. The Marburger Commission struggled to make sense of the issues, found it an "impossible mission," and was unable to move conclusively in one direction or another. After twelve days of meetings, four public hearings, and three subcommittee meetings on safety, economics, and plant operations, the chairman concluded that "no single policy can be derived unambiguously from the raw facts of the situation. On safety issues we have majority and minority judgments from the scientific and engineering communities. On economic issues we have diverse judgments from economists, lawyers, brokers, bankers and regulators." The one hundred or so pages of amplifying, dissenting, and clarifying opinions accompanying the two and a half pages of "General Conclusions" provided ample evidence of the differences. There was no disagreement, however, with the finding that "the Shoreham plant's long construction time and its staggering expense have contributed to a loss of public confidence on Long Island in traditional sources of judgment on utility planning and regulation."[51]

Given the divergent views, what could the Marburger Commission advise the governor with respect to safety and economics? Among other matters, it found that Suffolk County had adopted its position after commissioning studies of reasonable quality and that Suffolk's action was not frivolous. Since Suffolk County had not acted unreasonably, intervention by New York State was not warranted. The commission found that "there is an economic advantage, although possibly small, to operating the Shoreham plant versus not operating it," but that closing the plant would not result in a significant economic penalty. (Specifically, the commission said "that the projections of the incremental consequences of Shoreham operation versus abandonment are far less than the projected economic consequences of Shoreham regardless of whether it operates or not.") To Governor Cuomo's energy advisor, this was the panel's most startling conclusion—that dumping an almost completed Shoreham might not cost any more than operating it. "The Commission provided more than anything else economic ammunition. Made it easier for others to join in opposition to the Shoreham plant. Intuitively, it doesn't make a whole lot of sense, i.e., the notion that you can spend at that point three and a half billion dollars and that not to operate it saves you money or [is] . . . almost a wash."[52]

The panel stated, to no one's surprise, that the plant was located on a site that would have been "unacceptable" by today's standards. It cast doubts on LILCO's ability to manage the plant, should it ever operate, and expressed strong reservations about LILCO's ability to implement an emergency plan without Suffolk's cooperation. The Marburger Commission also made a suggestion that the federal decision makers failed to hear, or chose to ignore, that a low-power OL for testing the plant should not be issued before the off-site emergency planning issue was resolved. Finally, in a gratuitous understatement that must have received Governor Cuomo's warm agreement, the panel members stated: "the Shoreham plant will probably prove to have been a mistake in the sense that everyone might have been better off if the plant had never been built."[53]

## Shoreham: A "Motherhood" Issue

Between his early statement in support of Suffolk County and the Marburger Commission report, the governor was clearly impressed by the outpouring of opposition to the Shoreham power plant. The people he was encountering in Long Island were not a bunch of "crazy anti-nukes. They were suburban housewives, who were genuinely concerned about the effect that the plant might have on them. Whether that concern was based on sound evidence is almost irrelevant in one sense. They believed it to be true. In their minds it was true. Whether the scientists agreed or disagreed."[54]

Dr. Marburger had a similar reaction after listening to the public. He said that his experience as commission chairman profoundly changed the way he thought of the relationship between science and the public. "Suddenly I saw this in a nonscientific light. It didn't matter so much if the facts on evacuating the plant area were right or wrong. If it governs the way people behave, then the factual becomes almost irrelevant."[55]

The scientists' demeanor did not help them very much. At a public meeting on Shoreham, the governor is reported to have asked scientists from Brookhaven: "Where are you going to put the waste?" The response is described as "mumbo jumbo: 'There are scientific solutions. It's just the political system. You don't have to worry,' etc. What struck you," according to the governor's energy advisor, "was how incapable the scientists were of communicating with the normal citizen. They couldn't speak the language. It was almost intellectual contempt bred by overfamiliarity with an issue. *They* knew nuclear power. *They* knew how it worked. In their minds, there was no reason for people to fear."[56]

LILCO displayed a similar arrogance, and with the same results. The "quiet and self-contained" LILCO chairman, Charles Pierce, again stepped out of character in a speech to LILCO stockholders. He called his opponents' attacks "an abusive, thoughtless, slanted and repetitive barrage of rhetoric which may be unparalleled upon the American scene." Pierce characterized the Suffolk County legislature as "mesmerized and intractably dominated by a small tightly knit organized group opposed to operation of the plant." He also attacked "network pundits . . . would be investigative reporters . . . political hopefuls and . . . so-called consumer advocates."[57]

Pierce's intemperance may have been due, in part, to LILCO's worsening financial position. As Shoreham's cost rose to $3 billion at the end of 1982, for each month of delay $45 million in carrying charges were added to the cost. In March 1983, Moody downgraded LILCO's bond ratings to the lowest investment grade and reclassified its preferred stocks as a "speculative investment." Standard and Poor labeled LILCO's bonds as "speculative" soon after. In May 1983, Standard and Poor downgraded LILCO's preferred stock and bonds from BB plus to BB and BB minus, respectively. At about the same time, LILCO asked the New York PSC for a 56.5 percent rate increase to cover the cost of opening the plant and operating it through 1987. This was the largest request for an increase ever filed by a utility. Worry over rate increases heightened when the Grumman Corporation said it had postponed some expansion plans because of concern over future electricity rates.[58]

By June 1983, Shoreham was 99 percent complete and was undergoing final checks. Although *Newsday* likened it to "the feel of an airplane poised at the foot of the runway," a poll found that 52 percent of respondents believed the Shoreham plant should not be completed. The poll found that 76 percent would trust Governor Cuomo to tell the truth about Shoreham, and at this early date, 73 percent would trust the NRC to do likewise. A bare 50 percent said they would trust LILCO.[59] In 1983, the Shoreham issue dominated the campaign for county executive with the Republican incumbent opposing the opening of Shoreham. The Democratic candidate, Patrick Halpin, argued that Suffolk could not prevent the plant from opening and urged Suffolk to concentrate on making its operation as safe and economical as possible. Former County Executive Klein somewhat luridly likened the Shoreham issue to "riding a torpedo—it's hot, wild and fast."[60] After 1983, it was almost impossible to get elected to political office in Suffolk County without embracing an anti-Shoreham position.

The media was heavily involved too. In 1982, LILCO hired David Garth, a political consultant, to help it communicate with the public, and LILCO's chairman appeared regularly on local radio stations to praise Shoreham. LILCO arranged for the insertion of a twenty-four-page booklet describing Shoreham's history in weekend editions of *Newsday*. Shoreham supporters and scientists—organized into Citizens for an Orderly Energy Policy—received financial backing for their efforts to educate the public about Shoreham's worth from the U.S. Committee for Energy Awareness, a Washington-based pronuclear organization. Not to be outdone, SOC began a series of sixty-second radio spots that built on LILCO's lack of credibility. The spots raised sensitive issues such as "Can you believe LILCO in evacuation planning given its past blunders?"[61]

By now Shoreham's operation had ceased to be a technical question. It had become a political issue of importance "as a national symbol, both within and outside the nuclear community."[62] The political stakes had mounted rapidly since the early years when Shoreham had been opposed only by a small group of antinuclear activists. In 1982, the anti-Shoreham group was large and vocal enough to convince the entire Suffolk County legislature that fighting Shoreham was an issue worth championing. There was virtually no voice in government speaking for the plant. "By 1982, there wasn't a politician who felt any longer that he could take a stand in favor," said LILCO's then-president Wilfred Uhl. Matt Cordaro added: "Their motivation was the feeling that the public didn't want the plant. This is the way the politicians work; they read polls. They see LILCO is unpopular. They see that the public has fears about a plant. We [LILCO] tried over and over again to reach a compromise by the back door, side door, etc. You can't make concessions with someone who doesn't want the thing to happen."[63]

Any hope that the NRC might be able to quietly issue an OL for Shoreham without first conducting a public hearing had long since gone glimmering. When the NRC staff gave public notice that LILCO had applied for an OL, there were prompt requests for a hearing from persons "whose interest may be affected by the proceeding" and who were therefore entitled by Section 189 of the Atomic Energy Act to intervene in opposition. This time intervention did not come from a private group of lightly funded citizens such as the LHSG, who had fought the Shoreham construction permit. At the OL hearings, the parties opposing the license would be Suffolk County and the state of New York.

The Shoreham OL proceeding not surprisingly was destined to become "the most complicated, contorted NRC proceeding in history." Not even the Shoreham CP proceeding had prepared the NRC and its surrogate decision makers, the licensing boards, for "the second Shoreham war." The boards had adjudicated contested cases before but had never encountered opposition at the level found here. Herbert Brown, the county's lawyer, observed that "there is no area in the country . . . where political leaders are playing so major a role in opposition to licensing of a nuclear plant."[64] In this volatile environment, it would no longer be possible to conduct business as usual. Shoreham would be a test of the NRC boards' ability to adjust to a new situation, to use their expertise to respond to unprecedented public concerns. Would they win the trust of Suffolk residents in the competence and fairness of their decisions and satisfy public fears about the safety of the plant? If they could not, then any operating license the NRC might issue for Shoreham would appear to the public as just an act of federal government arrogance, not a meaningful resolution of the Shoreham controversy.

# 5 Governments in Collision
## Squabbling Over the Emergency Plan

During the next several years, there was prolonged and acrimonious sparring between LILCO and federal agencies on one side and the state and the county on the other about emergency planning. LILCO's OL application appeared to be in serious jeopardy when Governor Cuomo persisted in backing Suffolk County's opposition to the Shoreham plant. NRC's continuation of the licensing proceedings despite state and local refusal to participate in emergency planning confirmed the impression of Suffolk County residents that NRC was acting as an advocate for the utility.

**Emergency Planning, the Last Remaining Obstacle**

From 1983 to 1987, NRC's behavior continued to reflect the commission's strongly held view that the federal role was paramount in plant licensing and operation. From the agency's perspective, the question whether adequate emergency planning was possible at the Shoreham site should be determined solely by the NRC. If taken strictly as a legal matter this view was correct, for under the Atomic Energy Act all safety-related decisions relevant to the licensing of Shoreham were committed to the NRC. But the commission sometimes acted as though it believed no one else was entitled even to have an opinion. As the NRC saw it, for New York State and Suffolk County to decide on their own that no emergency plan could provide adequate assurance of safety if a nuclear accident occurred was presumptuous and disrespectful of the federal agency. The duty of the state and the county was to develop and be prepared to implement the best emergency plan they could; they should leave to the commission the decision whether or not the plan was adequate. By refusing even to offer a plan, and thereby blocking the smooth flow of the licensing process, state and county authorities were mounting an impermissible—and irresponsible—challenge to federal authority. The NRC thought the state and county governments had a duty to cooperate in bringing about the licensing of Shoreham, whether they favored it or not.

The NRC could perhaps counter state and county noncooperation by leaving the development of an emergency plan to LILCO itself, but this strategy left unanswered the problem of how the plan would be implemented in case an accident actually occurred. From the state and county perspective, assuring the safety of citizens in an emergency situation fell within the broad police powers assigned to their keeping by the federal system. These powers would be usurped if LILCO, a private corporation, tried to direct the sheltering and evacuation of Long Islanders after a nuclear accident.

The contrasting perspectives had far-reaching implications. Commissioner Victor Gilinsky pointed out that the Shoreham plant was widely viewed "as a bellwether for the moribund nuclear power industry. . . . The thing that has gotten people at the top of government going on Shoreham is the element of local veto that seems to threaten the authority of the federal government. There is the concern that it is something that could be duplicated around the country and cause havoc with the nuclear program."[1]

Although the commissioners believed that the Shoreham case would establish important precedents in federal-state relations, they nevertheless sought to remain independent of the political process and aloof from the controversy. Rather than discussing the emergency planning issue directly with the contending parties, the commission decided to give responsibility to FEMA—as the lead agency at the federal level for off-site planning—to identify the precise nature of the differences between New York, Suffolk, LILCO, and other participants and to see whether some kind of accommodation could be reached.

FEMA's effort was short-lived and unsuccessful. FEMA reported that it had contacted officials of New York and Suffolk and had offered assistance in holding a meeting but that neither had responded to the offer.[2] Notwithstanding the importance of the public policy concerns or the opposition of a governor of a large and important state, the commission itself made no further effort to hear the views of the parties in an informal setting. Henceforth, the issues would be discussed by independent licensing boards in courtroom-like settings with deference to legal niceties and highly structured, formal procedures.

In hindsight, this was not a wise approach to resolving the emergency planning controversy. It would have been practical, lawful, and productive for the commission to speak directly with the state and local people in meetings open to the public, to find out what was troubling them. Such meetings might not have resolved the problems but they would have clarified the points in dispute. Face-to-face encounters with Shoreham's opponents might also have made clear to the commissioners that they were headed for an unavoidable impasse in which the prospects for resolution were slim. As the former NRC general counsel observed: "There are some things that are best dealt with more efficiently and effectively at the commission level. When the governor of a state expresses strong views about something this agency has regulatory responsibility for in that state, that's one of the high priorities that this commission should be interested in and follow up on as effectively as it can."[3]

By 1983, the Shoreham plant was in the final stages of construction. LILCO was hoping to load fuel and begin operation by 1984. Both the NRC staff and the multiple NRC licensing boards working on the Shoreham OL application had already concluded that Shoreham's design and construction were essentially safe. (One board had been appointed in 1981 to conduct the OL hearing, which commenced in 1982. A second board was appointed in August 1982, to address physical security issues. And a third board was appointed in May 1983, to deal with off-site emergency planning issues.) All that remained was to review the adequacy of off-site emergency preparedness, but this proved to be a knotty problem.

The first issue of business concerned Suffolk County's proposal to terminate the licensing proceeding because of the absence of a local government emergency plan, as required by NRC regulations. The licensing board charged with emergency planning issues denied this request, calling it a clear attempt at "preemption in reverse," and concluded that an applicant for an operating license—in this case, LILCO—should be permitted an "opportunity to demonstrate that its plan could meet applicable regulatory standards."[4] Because of the importance and unprecedented nature of the decision, the licensing board referred its action to the commission's appeal board, which in turn passed it on to the commission itself for interlocutory review. Four of the five commissioners confirmed the board's action, stating that the NRC had not only the authority but the obligation to review a utility-sponsored plan in the absence of a state or locally approved plan. Commissioner Victor Gilinsky abstained, stating that the commission had failed to deal with "the actual issue in this case," which he characterized as "Can there be adequate emergency preparedness if neither the State nor the County governments will participate?" The answer, said Gilinsky, "is clearly no."[5]

Once again, the commission refrained from direct involvement, leaving to the licensing board responsibility for carrying out its "obligation," that is, for compiling a record on the major issues involved in the emergency planning controversy. The size of the evacuation zone was the critical agenda item. Suffolk County contended, among other things, that because of the county's population, demography, traffic patterns, and site specific circumstances, "an EPZ larger than 10 miles and perhaps as large as 20 miles is necessary" (see chapter 4). In clarifying this contention at a prehearing conference, the county explained that it was not seeking an EPZ of any particular size but merely "an *opportunity* to demonstrate the impact of local Long Island conditions on the appropriate size of the Shoreham EPZ."[6]

This particular contention had been heard before. The earlier Brenner Licensing Board (named for its chairman, Lawrence Brenner) had recommended that whether or not contentions were filed, the EPZ size issue was of sufficient importance to be raised and discussed by a licensing board in future proceedings.[7] Notwithstanding this advice, only one year later the newly constituted Laurenson Board (chaired by James A. Laurenson) agreed with the NRC staff and with LILCO that "this type of case-by-case attempt to litigate the extent of a plume EPZ" was an "impermissible challenge to NRC regulations." The board declined to hear evidence on the contention.[8]

Thus, the board charged with compiling an evidentiary record for the commission's review excluded from its deliberations the major sticking point in the emergency plans drawn up by Suffolk County. Because of the convolutions and delays in the various board proceedings dealing with the Shoreham OL, it took almost four years before the commission finally got around to reviewing the Laurenson Board's exclusion decision. Then in 1987 the commission upheld the board's finding that the intervenors' contentions could not be litigated in the Shoreham OL proceeding.[9]

In short, NRC regulations precluded Suffolk County from arguing at the hearings that the ten-mile-radius plume EPZ needed to be enlarged on safety grounds. As the NRC interpreted its regulations, the only acceptable basis for adjusting the EPZ would have been "administrative considerations," for example, to avoid the inconvenience of having the EPZ boundary run through the middle of a hospital or just nick the corner of some governmental jurisdiction. This kind of decision making—excluding a major safety issue from the hearing apparently because of a legal technicality—exposed the commission to criticism for appearing legalistic and arbitrary. Former Congressman Thomas J. Downey, a Democrat from Amityville in Suffolk County, remarked: "The more I watched the process unfold, the more skeptical I was of the sincerity with which they dealt with things like the shadow effect, evacuation, and a whole variety of other problems, because I believe they became advocates, not regulators."[10]

There is no doubt that the exclusion of Suffolk County's EPZ contention from the hearing did, in the short run anyway, help LILCO. It may not be fair, though, to accuse the commission of acting arbitrarily simply to advance LILCO's cause. In fact, the NRC's practice of excluding from adjudicatory hearings those issues that had previously been resolved in generic rule making was long established and had a sound basis in administrative law.

Under the Administrative Procedure Act, agencies generally are free to conduct their regulatory activities both by case-by-case adjudication and by rule making, as appropriate to the task at hand. In the regulation of nuclear power many issues arise that may reasonably be treated as "generic," that is, applicable to all or nearly all nuclear plants. They are appropriately dealt with once and for all by rule making rather than over and over again in successive individual adjudications. If the procedure were otherwise (that is, if litigants were generally allowed to challenge a generic rule when it is applied in a licensing proceeding), the agency could in effect be required to redo its rule making in each new adjudication. The whole point of rule making—improved efficiency and predictability of agency processes—would be lost.[11]

The ten-mile-radius EPZ rule that the commission relied on (some might say "hid behind") to exclude the intervenors' twenty-mile EPZ contention from the Shoreham OL proceeding was a generic rule intended to apply to emergency planning at all nuclear plants. The rule rested in part on general findings about likely radiation exposures from postulated accidents at nuclear plants in general. Particularly relevant was the observation, as summarized by the commission: "that the probability of large doses from core-melt accidents drops off substantially at about 10 miles from the reactor."[12] The purpose of the rule, as the commission interpreted it, was to specify an emergency planning zone around plants in which arrangements for "adequate protective measures" in case of a nuclear emergency must be set up in advance. The theory of the rule was that it was sufficient for the EPZ to include those areas in which prompt sheltering or evacuation would be needed to protect the population from radiation hazards. Farther away, there would be more time to devise and carry out on an ad hoc basis any additional protective actions that might be necessary in an actual emergency. Suffolk County's contention at the OL proceeding that Shoreham should have a twenty-mile EPZ was clearly at odds with this rule and the generic findings underlying it.

A closer question was whether the intervenors could argue for smaller adjustments in the EPZ, give or take a mile or so, in order to achieve "adequate protective measures." Not even this was allowable, said the commission. The adequacy of emergency planning was "not a precisely defined concept," and the EPZ rule did not contemplate such fine-tuning. Whatever one might think of such a rule, the commission's reliance on it to exclude the county's contentions was almost certainly legally sound.

Still, one may question whether, given the high level of controversy over Shoreham and the central importance of the emergency planning issue, the commission acted wisely when it refused to let the intervenors litigate their case for an enlarged EPZ. NRC procedures were sufficiently flexible to have allowed litigation of the county's contention, if the commission had really wanted to have the issue debated.[13] The commission seemed willing enough at other points in the OL proceeding to bend its rules, or even change them, when this helped remove obstacles to licensing Shoreham. But licensing would not have been facilitated by a public fight over the EPZ issue before the licensing board (although open consideration of the questions the county wanted to raise might have facilitated public acceptance of Shoreham later on), so it is not surprising that the commission did not allow it.

Avoiding unnecessary complications in the licensing process was certainly a reasonable goal for the commission. Former NRC Commissioner Peter Bradford has suggested, however, that the commission had additional, less creditable reasons for wanting to avoid discussion of radiation hazards beyond ten miles. Bradford, appointed by Jimmy Carter in 1977 as the youngest (and at six-feet-seven by far the tallest) commissioner up to that time, had been something of a maverick throughout his tenure on the NRC. Sometimes joined by commissioner Victor Gilinsky but often the lone dissenter to NRC decisions that smoothed the way for licensing plants and kept them in operation, Bradford acquired the reputation, unusual for an NRC commissioner, of being "antinuclear." This was a characterization he rejected, noting once in answer to a questioner that a better description of his views would be "anti-pronuclear." This "anti-pro" perspective stayed with Bradford when he later became chairman of the New York PSC, a position that would give him considerable influence over the ultimate fate of Shoreham. Looking back on the Shoreham OL proceeding conducted by his successors at the NRC, Bradford noted: "NRC could not give credence to the intervenors' claim that a twenty-mile EPZ could be considered. They couldn't live with the results. It meant that some past and/or future plants wouldn't pass the test. NRC would never face up to a standard that would disqualify an existing plant."[14]

Taken literally, Bradford's charge is a serious one, implying that even after Three Mile Island the commission preferred to risk inadequate emergency planning than to confront the possibility of having to shut down (or refuse to license) a nuclear plant. In fact, the NRC has not permanently disqualified any nuclear plant for emergency planning deficiencies. As the Shoreham proceeding continued to unfold, the commission took further actions of the kind that made Bradford's claim seem plausible.

## The Legal Authority Issue

Given the state and county refusal to help with emergency planning, LILCO became the first utility in the nation to draft its own emergency plan and rely on the services of its own employees to perform emergency functions. The state and county contended that LILCO lacked authority under state law to perform certain elements of its plan such as directing traffic, blocking roadways, activating the emergency sirens, or broadcasting emergency messages, and that consequently the plan could not be implemented.[15] When the licensing boards and the courts agreed, these concerns became an important issue in litigation before the courts of New York State.

Following extensive litigation, the New York Supreme Court found that under New York State law a private corporation such as LILCO could not perform certain functions that were inherently governmental in nature and that fell within the state's historic police power. Justice Geiler warned the state and local governments that they would be breaching their "fiduciary duty" to protect the welfare of their citizens "if they permitted a private corporation to usurp the police powers which were entrusted solely to them by the community." LILCO received another judicial blow one month later, this time in federal court, when a U.S. district court in New York ruled that the state and county could not be obligated under federal law to participate in emergency planning for Shoreham.[16]

The court decisions caused new problems for LILCO. Yet another NRC licensing board, the Margulies Board (named for Morton B. Margulies, who replaced Laurenson as chairman), concluded "that the LILCO plan is fatally defective . . . [in that] the applicant does not have legal authority to implement the plan. . . . [Further,] the opposition of the State and County has created a situation where at any given time it is not known whether the plan would be workable."[17] Hence, there was no reasonable assurance that adequate protective measures could be taken in the event of an emergency at Shoreham.

LILCO took its case to an NRC appeal board, arguing that (1) federal law preempted local and state law, (2) in an actual emergency, the state and local governments would cooperate to carry out LILCO's plan, and (3) LILCO's lack of authority to carry out the plan was immaterial since the public would be protected nevertheless. The appeal board rejected LILCO's arguments and affirmed the Margulies licensing board. Summing up, the appeal board said: "If the current state of the law frustrates LILCO by giving the state an eleventh hour veto over operation of the Shoreham reactor, the remedy lies in the legislative arena."[18] The "legal authority" issue continued to frustrate LILCO until 1987, when the commission voted to amend the emergency planning rule to bypass the issue.

### Bound and Determined to License

Despite the lack of an approved emergency plan, LILCO was moving ahead to secure a low-power license that would authorize testing and operation at power levels not to exceed 5 percent of full power. Given the substantial uncertainty about emergency planning, another licensing board recommended that, "as a matter of sound public policy," fuel loading and low-power testing should not be allowed lacking reasonable assurance that the reactor could ever be licensed to operate at full power. A majority of the commission disagreed.[19]

Two of the five commissioners—Victor Gilinsky and James K. Asselstine—found the commission's position "ill-advised." Commissioner Gilinsky stated that he was "at a loss to explain the Commission's decision . . . except as an effort to weigh the scales in favor of a full-power license before the present majority evaporates on June 30, 1983." (The reference to the "present majority" derived from the membership on the commission of John Ahearne, whose term expired on the night that the decision was reached.) Gilinsky added: "In actuality, the utility proceeds at the risk of the public. In light of the fundamental change in plant conditions which results from the irradiation of fuel, and of the associated substantial increases in the cost of maintenance and plant modification, the common sense and responsible view is that a low power license should not issue where there is substantial question that full power operation of the reactor will ever be permitted." This warning against contaminating the interior of the plant with radiation without assurance that the plant would fully operate was frequently repeated, but it did not deter a majority of the commissioners.[20]

Like the commission's decision to exclude contentions about EPZ size from the Shoreham hearing, the majority's refusal to rule out low-power operation of Shoreham looks myopic in retrospect. At the time it appeared defensible. The commission observed that the purpose of a low-power license is to allow testing to begin even though issues important to full-power operation remain unresolved. Experience had shown that the remaining issues eventually would be resolved favorably (to the commission's satisfaction, at least, which is exactly what later happened at Shoreham), so that a full-power license would in the end be granted. Getting low-power testing out of the way and dealing with any problems it reveals provides a major advantage: the fully licensed plant can begin generating electricity much sooner.[21]

This was the generic argument for creating a low-power license in the first place, and it is the justification the commission majority offered for refusing to exclude low-power operation at Shoreham, despite the uncertainty about emergency planning. When, later on, the commission actually issued Shoreham a low-power license, three of the most liberal, environmentally aware judges of the U.S. Court of Appeals for the D.C. Circuit—Skelley Wright, Pat Wald, and Harry Edwards—refused to enjoin low-power operation.[22] Of course, this did not mean the court necessarily thought the commission majority right and Gilinsky and Asselstine wrong. The court was simply unwilling to say that the majority's decision had been unlawful. In the end, though, like many other Shoreham-related decisions that could have gone either way, the low-power authorization turned out to be an extremely expensive error.

## Local Politics Versus National Security

The commission had no sooner agreed to keep open the possibility of issuing a low-power license for Shoreham than a major mishap occurred in the construction of the plant. In August 1983, cracks were discovered in the crankshafts of all three of LILCO's backup diesel generators. Since diesel generators are used to comply with NRC requirements for reserve power for coolant pumps and emergency shutdown systems in case of a primary power outage, their breakdown meant further delays in the licensing of the plant. Even though the faulty crankshafts appeared to owe their defects more to poor manufacturing than to inadequate maintenance, the incident was cited to support the opponents' charges that "LILCO management was too incompetent to design, build, and operate a reliable nuclear power plant."[23] Governor Cuomo's energy advisor complained: "[If] the utility couldn't even properly maintain diesel generators as a backup system, you had no confidence they were going to be able to maintain, with any degree of integrity, the complicated sophisticated system related to the nuclear aspect of the power plant." The diesel generator problems delayed the completion of the plant for one year and contributed to the impression that the plant was poorly built, of inferior quality, and therefore dangerous. They also contributed to a feeling of uneasiness in NRC when it was claimed that thirteen other plants had similar problems with their Delaval diesel generators and NRC had done nothing to "regulate."[24]

The diesel problems led to a management shakeup at LILCO. Chairman Charles R. Pierce announced his resignation and was replaced by William J. Catacosinos as chairman and chief executive officer. The resignations of Wilfred Uhl, president, and three vice presidents followed in short order. Catacosinos, "a millionaire entrepreneur" and "an activist crisis manager" who had been a LILCO director since 1978, assumed office at a time when LILCO's financial condition was worsening, with bankruptcy a real prospect. *Newsday* reported that bankruptcy would be unprecedented in the electric utility industry, with dire consequences for the local economy. Nobody knew for sure what would happen. Who would provide power? what would the rates be? and how would the credit markets react? The *New York Times* published a debate on LILCO's financial future, with two economists speculating on "What Happens When a Public Utility Goes Bankrupt?" Professor Robert Lekachman, Distinguished Professor at Lehman College, City University of New York, believed that the company was so ill-managed it "should be allowed to go broke . . . [and] take the fiscal consequences of its own past misdemeanors." Mark Luftig, vice president of Salomon Brothers, thought that "bankruptcy would be a dangerous thing"; in the long term, rates would go up, not only for LILCO subscribers, but for utility customers "all over the country."[25]

LILCO was $100 million in debt by the end of 1983 and in strong need of external capital. Just before Christmas, LILCO announced it would have to forgo future dividends, and its stock plummeted to the lowest level in nine years. Salomon Brothers called it the "financially weakest major U.S. utility," and Standard and Poor reduced LILCO's preferred stock four notches to CCC, the lowest rating for a company still paying preferred dividends. In March 1984, the new chairman announced a $100 million austerity program to help keep LILCO solvent. He eliminated 987 jobs (one-fifth of all LILCO workers), cut the wages of nonunion workers, and halted common stock dividends for the balance of the year. By April 1984, the company's annual report stated that LILCO could be in default on $500 million in bank loans; its stock closed at five and three-quarters, the lowest stock level in the company's history. The *New York Times* reported that the cost of the Shoreham plant had risen to nearly $5,100 per kilowatt of capacity, "the highest level for a nuclear plant in the U.S."[26]

At the same time that LILCO was displeasing investors, it did nothing to ingratiate itself with its customers. LILCO withheld its semi-annual tax payment in school, town, and county property taxes to protest the county's assessment on Shoreham, claiming that the county's opposition to the plant's operation was threatening to make the property "worthless." This move was termed by the deputy county executive as "economic blackmail."[27] LILCO might have replied that it was a form of "economic extortion" for the county to exact taxes on LILCO property while trying at the same time to render that property valueless. (Such an assertion by LILCO was upheld by the New York State Supreme Court some years later, resulting in a penalty of $83 million against the county, town, and school district.) In any event, county tax law at that time permitted a taxpayer to be in arrears for up to three years without penalty, other than interest on the unpaid taxes. Thus LILCO's withholding was entirely legal. It was an effective tactic, also, for the shortfall of tax revenues left the county with serious cash flow problems, in severe risk of losing its credit rating and going bankrupt itself. The county warned that it was facing the specter of at least a 25 percent tax increase and layoffs of up to three thousand employees.[28]

At the same time that LILCO chose not to pay its taxes and imposed an austerity program on its rank-and-file workers, LILCO's treatment of senior management, in some respects at least, fell short of exemplary self-denial. *Newsday* reported that the new "austerity" involved substantial increases over LILCO's historic compensation programs, including an agreement to pay the new chief executive officer $600,000 should he leave the company, in addition to his salary of $220,000 and $10,000 for "personal security." (He would also receive a pension of 60 percent of his salary, or $165,000 if he retired at age sixty-five.)[29]

Whether LILCO's concern for the comfort of its chief executive found favor with the average Long Islander facing ever-rising electric bills, the utility's popularity would soon drop below empty when Hurricane Gloria arrived in September 1985. The worst storm to hit Long Island in this century, Gloria damaged substations and transmission lines and left more than seven hundred thousand customers without service. One week after Gloria, a hundred thousand (8 percent of LILCO's customers) were still without power. What infuriated Long Island residents the most was the absence of LILCO President Catacosinos, who remained vacationing in Italy for about a week after the storm hit. Some of the public interpreted his absence as yet "another indication of LILCO's arrogant and unfeeling attitude toward its customers."[30]

LILCO's financial problems did not go unnoticed in Washington. The papers reported that the new LILCO chief executive was making frequent trips to the nation's capital to acquaint administration members with LILCO's financial problems and find ways to circumvent state and local opposition. Catacosinos's visits appeared to pay off. Secretary of Energy Donald P. Hodel, who became a prime mover in the Shoreham rescue effort, noted that "the breakdown of federal, state, local and private utility cooperation . . . posed a potential threat to the viability of the nuclear power industry" and recommended "federal actions to remedy this situation."[31]

Although President Reagan had stated in an October 1984 letter to Long Island Representative Carney, a Conservative Republican running for his fourth term in Congress, that the administration would not impose federal authority in the face of objections by state and local governments, the Department of Energy (DOE), FEMA, and the White House Office of Science and Technology became involved with LILCO in intense negotiations aimed at saving the Shoreham plant. President Reagan's science advisor, George A. Keyworth II, thought there might be an "erroneous perception" about the feasibility of the Shoreham evacuation plan, and DOE met with FEMA, LILCO, and White House officials to develop plans for a field test of LILCO's emergency plan "that can be somewhat credible." Hodel said he "simply could not stand by and permit the . . . non-operation of a plant on a technicality when it appears that the plant has been constructed adequately." (Governor Cuomo wrote Secretary Hodel that he did not consider the lack of an evacuation plan a "mere technicality.") John S. Herrington, Hodel's successor as DOE secretary, told the Nuclear Power Assembly that "Shoreham must open" and stated that "local politics can . . . [not] be allowed to fly against national security."[32]

## Integrity of the Regulatory Process

The NRC heard the message from LILCO and the administration. The chairman held a private meeting with the head of the licensing board panel and key members of the NRC staff on March 16, 1984, and was reported to "impress on them the need to accelerate the Shoreham decision and . . . authorize low-power operation." Commissioner Gilinsky objected to this procedure, stating: "The company had not yet applied for low power authorization. The Chairman did not inform the commission about this meeting until several days later, and did not provide the commission with important information about it until two weeks later." Several days later, the chairman circulated a memorandum to his fellow commissioners proposing measures to speed up licensing and reduce regulatory delays and expedite hearings for a low-power license for the Shoreham plant.[33]

Soon after, the commission appointed a special licensing board— the Miller Board (named for Marshall Miller, the presiding officer)— to consider the utility's proposal to load fuel and begin testing, despite the defects in the diesels. To resolve the question the board set an "expedited" hearing schedule that spanned the Passover and Easter holidays. The hearing was halted when a U.S. district court issued a temporary restraining order, finding that the overhasty schedule compromised the plaintiff's rights to due process.[34]

The chairman's personal involvement precipitated a nasty dispute before Congress and within the commission, with two commissioners publicly accusing the chairman of "irregularities" and "misconduct." Commissioner Asselstine stated that it would be reasonable to conclude from the chairman's actions that he "had abandoned his role as the ultimate judge in this proceeding and had instead become an ad-

vocate for a particular outcome in the case—the issuance of a low
power decision within an unreasonably short time period." In harsher
words, he accused the chairman of creating "an appearance of impropriety that was not likely to survive judicial review" and "undermining
the credibility of licensing hearings and the integrity of our entire regulatory program." In response, Chairman Palladino observed that he
wanted to ensure that NRC delays of a license would not, per se, be a
determining factor in leading to the abandonment of the Shoreham facility, and that he sought only to "bring some measure of efficiency and
expedition to this protracted licensing proceeding."[35] By this time,
however, it was clear that the chairman's efforts to gain "efficiency" in
licensing Shoreham had backfired; in addition, his credibility was diminished, the presumed "collegiality" of the NRC vanished, and the
NRC's impartiality and integrity lay ever more open to question.

Subsequently, Suffolk County and New York State requested that
the NRC chairman disqualify himself from further participation in the
OL proceedings, claiming that he had acted improperly by considering
the financial interests of LILCO over the interests of the state and
county. Denying any impropriety "either in reality or appearance," the
chairman refused. A majority of the commission, with Commissioner
Asselstine dissenting, also dismissed the suggestion that the staff monitor more closely than it normally would LILCO's operational readiness
in light of the company's financial difficulties.[36]

### Exemption from a Requirement

Despite the setbacks, the commission was still intent on issuing a
low-power license. At NRC's suggestion, LILCO requested an exemp-
tion from the requirement for an on-site emergency power source
(known as General Design Criteria 17) to begin low-power testing. In a
departure from usual practice, the Miller Board held that the requirement did not apply to fuel loading and low criticality testing.[37] The
board noted "the unusually heavy economic and financial hardships
facing the utility . . . [and] agreed with LILCO that the low power license would help it gain access to external capital."[38]

Early in 1985, the commission granted LILCO's request for an ex-
emption from the emergency power source requirement, permitting
LILCO to move ahead with the final steps of its low-power testing pro-

gram. Commissioner Asselstine again dissented, stating that "the Commission should [not] waive one of its safety requirements so that a licensee with financial problems can 'send a signal' to Wall Street."[39] After rounds of challenges and appeals—both internally within NRC and ultimately to the U.S. Court of Appeals for the D.C. Circuit—the commission issued a low-power license on July 3, 1985. The license permitted LILCO to start up a nuclear chain reaction in the reactor core, thereby causing radioactive contamination to the plant. Cleaning up and dismantling a contaminated Shoreham, should a full-power license be denied, could cost as much as half a billion dollars more than if the plant remained "clean." The "cost/benefit balance" had thus been given a significant tilt toward granting a full-power license. Commissioners Gilinsky and Asselstine dissented, arguing that the commission should perform an environmental evaluation of the issuance of a low-power license, including a cost-benefit balance.[40]

Public reaction was predictable. Kessel said: "It's a tragedy and NRC and LILCO are culpable. LILCO believed that by contaminating it, there would be no choice except to open it. Used it as a wild card in this." Governor Cuomo stated: "LILCO has for years been gambling with ratepayers' money and losing. They are doing it again with this decision. One would hope they would learn from their mistakes." And New York State's attorney added: "They should have understood that economics was never going to make the difference to New York when lives were concerned."[41] But former NRC commissioner Fred Bernthal argued:

If the commission gets into the game of not doing its job because of political opposition, then we're in trouble. . . . It is not in the commission's mandate to test the political winds, to try to decide how happy everyone is going to be with this. You try to make sure that the public's views are heard, that public interests are protected. That's part of the mandate. . . . Once the requirements for licensing have been met, then the commission is required by law to issue a license.

Bernthal was in effect saying that to take cognizance of the position of the governor, the county executive, or the local residents in licensing a plant was to go outside the commission's legal authority under the Atomic Energy Act. This line of reasoning was consonant with Denton's articulation of the staff view: "If it meets our regulations, which had been established properly through administrative proceedings, and that was the law of the land, that's all that the commission and the staff, as its agent, could insist on. And if it met that, we would issue a license."[42]

All this was true enough, but it slid past the uncomfortable fact that, at the time the commission issued the low-power license, there remained a major question whether the plant really could meet NRC regulations insofar as emergency planning was concerned. (It also ignored the commission's responsibility under NEPA to consider the environmental costs of prematurely issuing an OL.) In hindsight, permitting the plant to load fuel and "go critical" was a short-sighted action. It looked as though the commission was intent on giving LILCO some positive signal to ease the financial hardship that had arisen because of uncertainty over Shoreham. Eventually it would end up costing the taxpayers and the utility millions of dollars in extra cleanup costs when "speculation" that the plant might never generate electricity turned out to be correct. NRC's former general counsel admits that it would have been improper to issue the low-power license "if we were not confident at the time that the utility planned to do the job."[43] No one with decision-making authority at the NRC seemed to be considering the question, Under the circumstances, is this course of action in the public interest?

## The Benedict Arnold of Suffolk County

The low-power licensing decision had a ripple effect. Suffolk County Executive Cohalan said the decision "caught my attention more than anything else" as he suddenly reversed his opposition to the operation of the Shoreham plant. Cohalan had been under intense political and fiscal pressure from several sources: from DOE secretary John S. Herrington, who advised him of "the importance of Shoreham to the administration's drive to reduce dependence on imported oil"; from Lynn Nofziger, former aide to President Ronald Reagan who was working as a consultant to LILCO; and from local officials, because of LILCO's refusal to pay property taxes on the most expensive single piece of real estate in the county.[44]

Calling the plant "safer now than in 1982," Cohalan issued an executive order on May 30, 1985, directing certain county officials to review the Shoreham emergency plan and conduct an exercise with LILCO's emergency response organization. He replaced deputy county executive Frank Jones as "field general" of the county's efforts to block Shoreham and fired Kirkpatrick and Lockhart, the attorneys who led the legal fight against the plant. Almost immediately, LILCO said it would pay $130 million to the county in back taxes. The *New York Times* welcomed the shift, stating "Mr. Cohalan's new stance is certainly in the public interest."[45]

Not everyone was happy with Cohalan's about-face. Governor Cuomo continued to state that Long Island's geographical configuration and limited east–west roadways made evacuation "virtually impossible." He asked: "What's changed? Did the Island shrink? Are roads wider? Were bridges built?" And he responded, "none of that changed." Cuomo said that the state would be ready to fight Shoreham's operation even if Suffolk would not.[46] Nearly three out of five Long Islanders disapproved of Cohalan's switch, and a majority of the Suffolk County legislature went to court to enjoin the county executive from withdrawing the county's opposition. Several state court decisions upheld the view of the legislative majority that the county executive lacked the authority he claimed under state law to direct the county to participate in LILCO's plan.[47] The legislature promptly voted to rehire the law firm as its special counsel, and a discredited Cohalan issued a second executive order rescinding the earlier one.

By 1985, Republican polls showed more than 65 percent of likely voters opposed to Shoreham's opening. Anti-LILCO sentiment appeared to be a key factor in many of the races for the state legislature and for Congress, as well as for the Suffolk County legislature. In 1986, a longtime supporter of Shoreham in Congress from eastern Long Island dropped out of the race as opposition to the Shoreham plant became the key test for Long Island voters.

## Fantasy Exercises

Before LILCO could gain a full-power license, it had to conduct a successful drill of its emergency evacuation plan for residents of the plant's neighborhood. The drill was to be conducted under the supervision of FEMA, which would also evaluate the results. Despite the uncertainties of state and local participation and LILCO's lack of legal authority to perform the public functions traditionally reserved to the states and localities, three out of five NRC commissioners urged FEMA to test as much of LILCO's plan as was feasible. This time, even the supposedly pro-Shoreham Chairman Nunzio Palladino joined Commissioner James Asselstine in questioning the usefulness of an exercise in which LILCO would merely simulate the roles of state and local officials.[48]

FEMA questioned the value of the exercise, too, stating it would be "dramatically different than is typical at other sites" and would not allow "a finding of reasonable assurance" regarding the adequacy of LILCO's emergency plan. Nevertheless, when pressed, they agreed to move ahead. Congressman Tom Downey commented that the NRC was searching for a way to circumvent local opposition and license the Shoreham plant: "Despite their own regulations, they appear bent on getting it open." Governor Cuomo called it "illegal, wasteful and futile . . . [and] an affront to the sovereignty of New York State." The two New York State senators were similarly outspoken. Senator D'Amato called it "illegal and a violation of states' rights" and appealed directly to President Reagan to cancel it. Senator Moynihan called it "ill conceived, . . . a waste of public funds and prima facie evidence of elemental incompetence" on the part of federal regulators. (To Suffolk County's Frank Jones, the exercise made no sense; it would be like "simulating a kiss from your girl friend from ten feet away.")[49]

Suffolk County was not happy, either. Asserting that persons playing the role of county officials in an emergency exercise would be usurping county police powers, the county passed a local ordinance that would subject any federal or LILCO official who participated in the exercise to local criminal prosecution. Even apart from this law's questionable rationale (under which role-playing by high school students at a student congress would be a "usurpation" of legislative authority), the law arguably intruded on the regulation of nuclear safety, a federally preempted area. This intrusion placed Suffolk County on a constitutional collision course with the federal government. A federal district court preliminarily enjoined the county from enforcing it, stating that the county's motive was "to continue its opposition to the Shoreham facility on the basis of a perceived radiological hazard." Thus the law "impermissibly intruded into a sphere of authority reserved exclusively to the federal government by Congress."[50] In short, the court saw Suffolk County, not LILCO, as the usurper.

With the threat of prosecution lifted, the exercise went forward on February 13, 1986, and involved approximately eighteen hundred LILCO employees and one hundred federal government participants playing the part of local and state personnel. LILCO assigned workers to drive buses and tow trucks, direct traffic, and monitor radiation of the evacuees. For the public, the only evidence that a drill was going on was the stationing of LILCO personnel and vehicles at some key intersections in the evacuation zone. State and county authorities called the test "meaningless . . . showing that LILCO employees can use touch-tone phones."[51]

The test was unsuccessful. FEMA's regional director Frank Petrone charged that there was pressure from Washington to weaken a report that was critical of the drill; he resigned his position. An NRC licensing board held that the exercise demonstrated "fundamental flaws" in the Shoreham emergency plan and did not meet NRC's regulatory requirements.[52] NRC scheduled a second exercise in June 1988.

The commission's insistence on holding the drills in the face of non-cooperation by the state and local authorities exacerbated local sentiment against NRC and lent credence to the notion that NRC wished only to license the plant rather than to assure the safety of Suffolk residents. State and local officials and the public were outraged that the utility would spend millions of dollars to test a plant that might never operate commercially. Looking back on the episode, NRC's general counsel William Parler questioned whether the licensing process should have continued: "If there is a situation where cooperation of state and local government is needed . . . for this agency to make its safety findings in emergency planning, why didn't the regulatory agency just stop the proceeding and force the decision to be made one way or other as early as possible, without having it drag on and on?"[53] But the commission would in fact soon declare that state and local cooperation was not necessarily "needed" for adequate emergency preparedness, or if it was, it could simply be assumed. This assumption became known as the "realism" doctrine.

## "Gutting" the Emergency Planning Rule

Confronted with LILCO's poor showing on the exercise, NRC amended its emergency planning rules in 1987 to incorporate the "re-

alism" doctrine, a presumption that state and local government officials would do their best to protect health and safety in an actual emergency, regardless of their earlier statements of noncooperation. The
commission noted there were "significant policy questions of equity
and fairness" when governmental noncooperation blocked operation
of a substantially completed plant. The statement continued: "It is only
reasonable to suppose that in the event of a radiological emergency,
state and local officials, in the absence of a state and local radiological
emergency plan approved by state and local governments, will either
look to the utility and its plan for guidance or will follow some other
plan that exists."[54] With the realism doctrine in hand, NRC could
henceforth license a nuclear plant in reliance on an emergency plan
prepared by a utility without state or local participation, on the "reasonable" supposition that the authorities would implement that particular plan, and do so effectively, if an accident actually should happen.

Hope and expectation thus took the place of missing hard evidence
on emergency preparedness. Many perceived the amended rule to be a
lowering of nuclear safety requirements. Commissioner Asselstine dissented from the change, stating that it undermined the very foundation
upon which emergency planning was based. For him, the "new emergency planning philosophy was nothing more than the Commission's
pre-1980 philosophy in new trappings. . . . Obviously, the Commission's commitment to emergency planning only lasts as long as it did
not get in the way of expeditious licensing of plants."[55] Although the
commission denied that the new rule was directed at any particular
plants, it was generally recognized that it would remove the biggest obstacles to starting up two controversial plants, Shoreham in New York
and Seabrook in New Hampshire, where emergency planning was also
a sticking point.

Public debate on the rule change was unprecedented. The rule elicited eleven thousand comments from the public, a record for NRC proceedings. So many officials wished to testify that the NRC was forced
to consult the State Department's protocol office for the proper order
of appearances for members of Congress and the governors. Four governors, three senators, and six House members appeared in person to
oppose the rule change. Governor Cuomo called it "a blatant political

fix" that flew in the face of traditional concepts of the state's power to protect the well-being of its citizens. He charged the commission with putting the interests of the nuclear industry and its investors ahead of public safety. Former Commissioner Bradford, who had become chairman of Maine's Public Utility Commission, called it "a serious mistake . . . [that would] only further disenchant the public with nuclear power."[56]

The amended rule brought to the forefront the long-standing struggle between the federal government and the states over nuclear safety. The NRC, concerned that the states were playing too large a role in emergency planning, was determined not to share authority in a decision that affected the "national interest."[57] Thus, the commission refrained from speaking with the governor and the other parties to the controversy because this might be interpreted as a threat to its independence of action. It also refrained from taking seriously the governor's word on emergency planning, and stopping the proceeding until the issue was resolved, because this too could be regarded as a de facto sharing of authority. To assert and maintain exclusive hegemony over all aspects of emergency planning, the commission seemed willing to risk appearing as a proponent of utility interests rather than an impartial adjudicator.

The amending of the emergency planning rule also reflected the frustration level of the commissioners. Here they had been appointed by the president to be the congressionally mandated arbiters of nuclear safety, only to be stymied in performing their legal function by a determined governor and a single local government. The commissioners saw their role as one of rendering a judgment on safety, and when they believed a plant was safe they could make things happen. They saw no reason to deny a low-power license simply because there were people opposed to it. And the state kept getting in their way. Governor Cuomo's energy advisor conceded that many NRC officials believed "this state is wrong; this state is just being obstructionist." The NRC technical staff were frustrated too. An NRC attorney explained: "They thought New York and Suffolk County . . . were a bunch of politicians. And you can't blame them. Engineers are trained to take problems and solve them. They don't understand the way politicians and lawyers operate."[58]

LILCO was frustrated too, of course, but for altogether different reasons. LILCO saw NRC as catering to the intervenors. To LILCO's president: "NRC leaned over backwards to give the opponents of the project every possible chance; to introduce what we thought was all sorts of irrelevant material." LILCO's vice president concurred: "The perception within the company was that NRC had to do something with its procedures to prevent endless hearings. . . . It took forever to get NRC to act on things. . . . It was unnecessarily slow." And LILCO's principal lawyer believed: "NRC was unable to figure out what the rules were on behalf of effective procedural constraints on evidentiary hearings. NRC tried to establish that it really was fair by giving procedural opportunities for people to oppose the staff and the applicant in ways that were relentless, excessive, and almost impossible to deal with."[59]

But New York and Suffolk did not think the NRC was giving Shoreham's opponents "excessive" opportunities, and neither did the public. Governor Cuomo's advisor stated that the "impression we had of the NRC was that . . . they had long since abandoned their congressional mission of regulating; instead they were promoting nuclear power. They were falling into the same trap as the old AEC. Nobody believed that the NRC was really there trying to protect the public." Tom Downey, a Suffolk representative in Congress, called the commissioners "plutonium producers of the first order. They approach the process by saying nuclear power is safe; that there is more ambient radiation in the atmosphere that we should be worried about than that which is generated by nuclear power. Nobody believes these arguments and certainly not the people of Long Island while the Shoreham dispute was happening."[60]

With both sides condemning the NRC, one has to suspect that the commission was doing something right, even if this would be hard to identify. In some cases the commission was blamed for matters over which it had little control. For example, revelations of ex parte meetings between LILCO, DOE, and members of the administration were looked upon as "another instance of NRC lack of fairness." Internal memos were obtained showing pressure being brought by DOE, with the blessing of the administration, urging NRC to issue the license. There was a feeling by the public that "NRC was part of this process." Even LILCO's attorney agreed that "NRC had no credibility. . . . There was hostility from New York and Suffolk County. NRC hoped it would go away."[61]

The hearing boards were no better liked than the commission. They were seen as an extension of the commission, "simply interested in putting the plant on-line." When the NRC in 1986 removed two of the three judges on a Shoreham licensing panel who had listened firsthand to the concerns and fears of those opposed to the plant and created another panel to take account of the "realism" argument, this just confirmed the public's belief that the NRC was stacking the hearings in favor of LILCO. New York's attorney said that the hearings were "really a farce. Every time they had a hearing officer who ruled fairly, they'd change him and put in someone who was more of a hatchet man. Discredit to the system of administrative law judges."[62]

The Chernobyl accident in 1986 confirmed fears that nuclear catastrophe was a realistic threat, not merely the remote theoretical possibility acknowledged by federal regulators. The NRC's director of nuclear reactor regulation admitted publicly that the disaster "might well complicate domestic policy" with regard to the licensing of new U.S. nuclear power plants. New York's attorney was more outspoken: "Chernobyl made everyone conscious that New York was not just a bunch of ranting idiots. . . . [The accident] gave us credibility."[63] At this point, if not before, it became understandable how a growing number of entirely reasonable people, including responsible state and local officials, could sincerely believe Shoreham could be dangerous. Also, they could reasonably conclude that the decision whether to let Shoreham operate should not be left to federal regulators who apparently thought emergency planning was safety overkill, something formally required by the regulations that would never really be needed.

# 6  Who Decides?

## State Economic Power

---

N RC's amending of the emergency planning rule came to naught. In the late 1980s, the focus of the Shoreham controversy shifted from the NRC licensing proceeding to the governor and the New York PSC. The NRC had exclusive authority to declare Shoreham "safe"; but once the state authorities decided that Shoreham should never operate, they had the power over LILCO's economics to force this result.

### Converging Pressures on LILCO

*The Public Takeover Movement*

As the NRC changed its rules in the effort to license Shoreham, New York and Suffolk hardened their opposition. If the two governments could no longer stop the plant from operating by the emergency planning route, they would use other means to achieve the same result. Governor Cuomo took the lead in employing the state's economic power to compel LILCO to accede to his wishes. "The Long Island Power Authority [LIPA] was created out of total anger towards LILCO," the executive director of New York's Consumer Protection Board (CPB) has said.[1]

In July 1986, dissatisfaction with LILCO among Long Island residents and their state representatives manifested itself in a move by the state legislature to create a public authority that would buy LILCO stock and transfer ownership of LILCO from current shareholders to a government entity. (This move followed closely on the heels of a state court opinion by the appellate division of the state supreme court, barring Suffolk County from establishing its own municipal development agency to sell up to $7.3 billion in tax-exempt bonds and buy a controlling interest in LILCO stock.)[2] The groundswell for a public agency to replace LILCO started in Suffolk County with the support of 70 percent of Long Island residents and was eventually supported by all three Albany power bases—Governor Cuomo, the Republican-controlled State Senate, and the Democratic-controlled Assembly—as a way to provide less expensive electricity to Long Islanders. How much would be actually saved was uncertain. There was no firm evidence that the state could operate the utility at less cost than LILCO, but it was believed that a major source of savings would come from the reduced cost of financing that a public power authority would enjoy and from eliminating the federal income tax that is included in the customer's electric bill.[3] In effect, Long Island power-users would be saving money by giving themselves a federal tax cut.

The bill provided LIPA with authority to begin negotiations with LILCO's board of directors as of January 15, 1987, concerning a friendly takeover whereby LIPA would purchase LILCO at a price of up to $18 a share of LILCO stock, with the proceeds of the sale of tax-exempt municipal bonds. If LILCO did not accept the offer, the new authority was empowered to bypass the board and attempt a hostile takeover through stock purchases. (As the Sawhill Commission described it: "the primary difference between a negotiated acquisition and a tender offer is how the requisite shareholder approval is obtained.")[4]

Under either a "negotiated" or a "hostile" tender offer approach, the price to be offered by LIPA would be determined by a valuation of the stock, rather than by appraising, brick by brick, the value of LILCO's assets. Irving Like, a member of the LIPA board, pointed out that there was no other statute in the United States like the one that created LIPA and empowered it to take over LILCO. Rather than condemning and purchasing the assets of a utility company (as such takeovers were usually effected), the statute set forth in quite comprehensive detail the machinery and the procedures for buying out LILCO stockholders at current market prices. "The advantage," said Like, "is that the issue of evaluation is obviated because the stock is what it's trading for on Wall Street, on the Exchange."[5] Although Long Islanders were optimistic that the state takeover would result in a lowering of electric rates, the *New York Times* disapproved, calling it

the same old game. . . . Long Island ratepayers will eventually have to decide how much they are willing to pay to keep Shoreham idle. New York State must decide whether all its taxpayers should subsidize Long Island's resistance to nuclear power [that is, in the event that the interest costs of other state tax exempt agencies were to rise]. And Washington must decide whether to let states and localities veto the nation's development of nuclear energy.[6]

The legislation also required an abandonment of Shoreham and a determination by LIPA that rate savings would occur. A companion measure, passed at the same time, prohibited LILCO from recovering any of the cost of Shoreham from the ratepayers unless the plant was "used and useful" and put into operation by the end of 1988. LILCO challenged this provision of the bill in federal court because it applied only to LILCO among the state's private utilities. The provision was declared unconstitutional by federal district court judge Howard Munson in August 1987, as a violation of the equal protection clause of the U.S. Constitution.[7]

Although Governor Cuomo supported a public takeover, it was uncertain whether he was comfortable with the notion of a publicly run utility "that raised complex questions about the relationship between government and private enterprise." His senior aides said that his real goal was to use the legislation "as a club to force the utility to negotiate over the future of Shoreham."[8] As was his wont when faced with difficult decisions, Cuomo appointed a commission to study the matter. He picked John Sawhill, a former federal energy administrator, as chairman to investigate the potential for cost saving.

The Sawhill Commission found that the replacement of LILCO by LIPA might, under certain conditions, save ratepayers "in the range of 7–9 percent" over a fifteen-year period. "These savings would arise primarily from the absence of federal income taxes for LIPA" and from other financial advantages (of lesser importance) associated with interest rates and capital structure. The finding was tentative; the advantage or disadvantage of going to public ownership and/or operating Shoreham, as the Sawhill panel looked at it, was very much an artifact of the federal and state tax systems and the financial markets. The Sawhill Commission concluded that the "preferred" alternative to a public takeover was a "negotiated agreement" with LILCO's management and board of directors leading to a stock purchase.[9]

The investment banking firm Lazard Freres and Company also studied the cost-saving issue for LIPA. It concluded that Long Island residents and businesses would save nearly 12 percent in electricity costs over fifteen years under public ownership, because LIPA would not be required to pay income tax on its earnings and could secure financing at lower interest rates because of another tax break (its bonds would be tax-exempt).[10] Some other taxpayers would have to take up the slack created by these "savings," but this was not the concern of Long Island.

Almost two years passed before LIPA was prepared to take the necessary steps to win control of the utility. On March 30, 1988, encouraged by the Lazard Freres study, LIPA made a friendly low-ball offer of $8.75 a share for each of LILCO's 110 million shares of common stock (in one of the largest public takeover efforts of any private company) and was turned down by the utility. The stock was then selling at $9.125 and LILCO called the offer a "firesale price." Since state officials were negotiating with LILCO executives at the same time to reach agreement on a plan whereby LILCO would abandon Shoreham and remain investor-owned, the buyout offer was viewed by some as a ploy to bring LILCO to the negotiating table.[11] When LIPA increased its bid by $137 million one month later (raising its offer for LILCO's common shares to $10) and then voted to wage a proxy battle for control of the utility, negotiations between LILCO and the state became more spirited.

It is certain that LIPA's hostile proxy effort galvanized LILCO into thinking more seriously about the abandonment of Shoreham. At the time, LILCO was facing bankruptcy as well as tremendous opposition to opening Shoreham. "One way to resolve its problems," said LILCO's vice president Joseph W. McDonnell, "was to sell the entire company to the LIPA. [However,] they never came forward with a serious offer and it became evident that was not the solution that the governor desired." Like, then a member of the LIPA board, agreed that the governor's opposition was a critical factor in dooming a forcible takeover:

The five representatives he controlled on the LIPA board would never let the board authorize an offer of more than $12 a share. This was at a time when the investment banking consultants were telling LIPA that, if you want to be successful, you had to offer about $13 or $14 a share. And the board split and the four of us who were in the minority . . . couldn't muster a majority. . . . At the same time . . . the governor announced that he and the chairman of the [LIL-CO] board had reached a settlement.[12]

Another LILCO vice president confirmed that "takeover was a threat; it was a major public issue. . . . The creation of LIPA opened the door for some discussions with the state and ultimately led to the settlement." Charles Pierce, LILCO's former chairman and chief executive officer, believed that LILCO's eventual agreement to abandon Shoreham resulted directly from efforts that were launched under the LIPA legislation. "Had it not been for the public power constituency and the LIPA statute," said Like, "Shoreham would be operating today."[13]

*PSC Transformed from Rubber Stamp to Antinuclear Avenger*

The New York PSC, which gained new visibility as an arbiter of rate increases for the Shoreham plant, has elicited sharply divergent reactions from those familiar with its activities in regulating electric utilities. Created by the state legislature in 1907 to ensure the availability of safe, adequate utility services to New York consumers at reasonable rates, the New York PSC was given legal authority in 1921 to regulate the rates, financing, safety standards, and quality of service provided by utilities in the state. Characterized as "one of the most active and sophisticated of the state regulatory bodies," it gained an ambiguous early reputation. Consumer-oriented observers thought it permitted the utilities to charge higher rates than were necessary and allowed an excessive rate of return. They also considered it lax in supervising the utilities' operating practices and maintenance procedures, permitting inefficient practices to continue to plague the industry.[14]

Rate making is at the heart of the PSC operation. Under New York State law, utility rates must be "just and reasonable," that is, "they must be adequate to cover the utilities' operating costs and provide an adequate return to encourage further investment in a capital-intensive industry." Although New York generally followed the rule that a power plant had to be "used and useful" (that is, operating) before its cost could be added to rates, New York law allowed a utility to recover construction costs of a project, even an abandoned one, if they were prudently incurred.[15]

The PSC was an early booster of nuclear energy and the construction of nuclear power plants. Notwithstanding the "prudency" decision in 1985, which disallowed $1.4 billion of Shoreham's cost from LILCO's rate base because of mismanaged construction, the PSC generally supported LILCO, acting favorably on its continuing requests for rate increases to enable the utility to survive financially. From 1970 to 1981, for example, the PSC granted LILCO thirteen rate increases, totaling $521.2 million. (At the same time, it rejected a petition by Shoreham opponents that the project be abandoned, noting that too much money had already been spent by LILCO on the project.) More increases followed. In 1984, the PSC granted LILCO a record $245 million increase at a time when financial strain brought on by the Shoreham project was driving the utility toward bankruptcy. The increase helped convince the banks to enter into a revolving credit agreement that kept the company afloat financially.[16]

At the start of 1986, in a four-to-three vote to help LILCO "regain its financial health," the commissioners granted LILCO $68.7 million and accepted guidelines that would allow LILCO to receive rate increases of 5 percent a year for the fifteen years following the opening of Shoreham. (The phase-in plan assumed that Shoreham would go into commercial operation in January 1987.) The *New York Times* estimated that rate increases of this magnitude, compounded over the fifteen-year period, would have caused Long Island electric rates to double.[17]

It seems clear that the PSC's continuing approval of rate increases encouraged LILCO in its quest for an OL for Shoreham. Richard Kessel, then head of New York's CPB, held the PSC as "responsible as LILCO for what happened. Shoreham was not supposed to be part of the rate structure. But every time LILCO went to plead its case, it threatened bankruptcy." Former PSC commissioner Karen Burstein agreed. In the early years, she says, the PSC allowed LILCO to be the most "wasteful, careless, self-regarding entity. . . . They were imposing no financial discipline on LILCO; the thing was completely out of control. Inefficiency was being rewarded. The PSC insulated them from risk." The former chairman of the Grumman Corporation was similarly critical: "The PSC seems to have forgotten the word 'public' in their title. . . . They act as if they are in business solely to supply utilities with cash on demand."[18]

Taylor Reveley, LILCO's lawyer, acknowledged the PSC's willingness to do whatever seemed necessary to keep LILCO afloat:

The PSC . . . thought the plant ought to be built. It generally thought it was costing a lot of money and taking a long time to get finished. The best thing to do was to finish. And it tried, while beating on LILCO pretty relentlessly in the prudence hearing (War Crimes trial, we always called it), to keep the company going also. And one way to do that was to put in the rate base some of the cost. . . . The PSC could have pulled the rug out from under them by refusing any increase.[19]

## The "Death Blow" for Shoreham

By February 1986, the PSC had already included $1.8 billion of Shoreham's cost in LILCO's rate base to help the utility avert bankruptcy. With the public clamoring for action to hold down LILCO's rates and the public takeover issue coming to a head, the governor said that if he "could start all over" with the PSC, there would be a different commission.[20] True to his word, the governor ousted chairman Paul Gioia, whose term as chairman had not yet expired, and replaced him with Peter A. Bradford, the consumer-oriented chairman of the Maine Public Utilities Commission and former member of the federal NRC.

Close observers offer various explanations for Cuomo's reluctance to act at an earlier time. One explained: "Gioia postponed bankruptcy for the governor. Bankruptcy would have been politically embarrassing. As long as Gioia wasn't called for what he was doing, he could keep on doing it for the governor and postponing the problem. When it became a matter of public issue, the game was over." The former president of LILCO defended the governor's prolonged refusal to interfere with Gioia's indulgence of LILCO, stating: "It would have been an irresponsible act for him to permit bankruptcy. It would have made it more difficult for every utility in New York State to borrow money; higher interest rates, etc." Another LILCO official explained that "the PSC is required to provide the company with the rates it needs to survive as a going concern. In bankruptcy, the PSC loses control over the ability to set rates. The courts can set the rates. So that made it more of a danger." New York's CPB confirmed this view, finding bankruptcy as an approach to be "fraught with uncertainty. . . . It stands to reason that rates set by the Bankruptcy Court, whose primary concern is protecting creditors, would substantially exceed the rates to be expected under conventional regulation."[21]

The PSC's policy of authorizing cash flow infusions to finance Shoreham's construction was reversed in 1987. The reconstituted PSC—with five new members out of seven—adopted a tough stance toward LILCO, stating that a continuation of financial relief would only prolong a "financially debilitating impasse." The PSC urged the parties to "break the long deadlock so that the Long Island community can look to the future with assurance that their power supply will be safe, adequate and reasonably priced." In a unanimous ruling, the PSC voted to deny LILCO's request for an $83 million rate increase and told the utility and the state to negotiate an end to the Shoreham situation without reliance on plant operation. The PSC's shift from a pro- to an anti-Shoreham policy exerted strong financial pressure on LILCO to give up on the plant. Plant opponents called the PSC's turnabout a "historic decision" and "the death knell" for Shoreham. Governor Cuomo called it "a victory for the people of Long Island." Investment bankers called the decision "extremely negative," and company officials said it could hurt LILCO's borrowing capacity on Wall Street.[22]

The PSC's rejection of LILCO's rate increase provided added incentive to LILCO to move simultaneously in two diametrically opposed directions: to proceed with its effort to gain a license for the plant so that most of the costs could be charged to consumers, and to negotiate an end to the fight over Shoreham. As pressures for a settlement mounted from influential businessmen, politicians, state utility regulators, and LILCO shareholders, LIPA and LILCO met together for the first time. Daniel Scotto, the utility-bond analyst for L. F. Rothschild and Company, said: "The handwriting is on the wall. It's time for Mr. Catacosinos to give up the plant and save the company."[23]

### Settlement, to Save the Company, Not the Plant

Governor Cuomo reached an "agreement in principle" with LILCO in May 1988, to sell Shoreham to the LIPA for $1, which would decommission the plant at LILCO's expense. In exchange, LILCO would receive significant rate increases—three years of annual 5 percent rate increases and an additional seven years of annual "target increases" of 5 percent—and would resume payment of dividends to LILCO's shareholders beginning in 1989. LILCO was expected to take a $2.5 billion tax write-off for the loss of Shoreham, "allowing the company to pay no federal taxes for at least a decade." (This part of the agreement was approved by the Internal Revenue Service [IRS] in September 1988.)[24] But LILCO would nevertheless recover in rates much of the cost for building the abandoned plant through the creation of a fictitious "regulatory asset" equal to part of what LILCO had expended building Shoreham. This "asset" would go into LILCO's rate base and enable LILCO to charge higher rates. As Irving Like described it:

It was in their own financial interest to settle. By giving up Shoreham, they were given what is called a "financial regulatory asset"; a financial asset was credited on their books in the amount of $3.66 billion of their investment in Shoreham. It was a risk-free asset, on which they were guaranteed a return over the period of the settlement for about fifteen years, and they were guaranteed rate increases for a certain number of years. So, if you were the company, and you could have the equivalent of Shoreham in your rate base without the risk of operating Shoreham, . . . you're much better off simply taking the financial asset and not having all the headaches.[25]

The deal involved authorizing legislation from the state legislature, which would give the New York Power Authority (NYPA) responsibility for building gas-fired turbine power plants on Long Island to replace the 800 megawatts Shoreham would have generated. Governor Cuomo said, "We have gotten everything we wanted. Shoreham has to die." Susan Benkelman of *Newsday* termed it a "setback for nuclear power in the U.S., for in almost every way—cost, lack of public acceptance, construction problems and delays—Shoreham embodied the failures of the industry, which has not seen a new reactor ordered since 1978." *New York Times* correspondent Matthew L. Wald saw it differently, saying that the harm to the industry had long since been done; the "significance to the nuclear industry of Shoreham's cancellation is like having a dying horse break a leg."[26]

Although the deal was overwhelmingly supported by LILCO shareholders and by Wall Street, Suffolk County legislators had second thoughts about returning LILCO to financial health, calling it "economic blackmail."[27] LILCO customers were already paying the second-highest electric rates in the United States based on 1987 industrywide figures, behind only Consolidated Edison Company of New York. The Long Island Association, a group of Long Island businessmen, was similarly opposed, and Grumman Corporation claimed that the rate increases would lead to layoffs and a significant drop in local economic growth. The state legislature also found the agreement too expensive for the utility's customers; in a major political defeat for the governor, they withheld approval. Frank Lynn wrote in the *New York Times:* "Long Island came of age politically yesterday . . . emerging as the 'single most powerful delegation' in the state legislature." Governor Cuomo accused the legislature of endangering the lives of the 2.7 million residents of Long Island by refusing to accept his plan, saying that "the legislature was playing Russian roulette. They have now pulled the trigger." LILCO said it would remain agreeable to the settlement (as long as it had not received NRC's permission to operate the plant), but it would still continue to press forward to gain a license.[28]

## Neither Capitulation nor Extortion

A rather unseemly race between the federal government and the state was now in full force. "The perception," wrote the *New York Times,* "is of a high-stakes race between the regulators in Washington, fighting to open the plant as quickly as they can, and the Cuomo administration, struggling to push through its plan to close the 809 megawatt plant." New York's bargaining chips resided in its power to force LILCO into bankruptcy. This was countered by LILCO's ability to operate Shoreham once NRC granted a license. The latter was still a possibility. In September 1988, the U.S Court of Appeals for the First Circuit had affirmed the NRC's new emergency planning rule allowing utilities to use their own evacuation plans when local and state governments were noncooperative, and FEMA had given LILCO positive grades for its most recent evacuation drill. LILCO said publicly that "it is now whichever comes first, a license or the settlement."[29]

It was unlikely, however, that either side would make good on its threats. For its part, New York could not have been eager either to ruin LILCO financially or to carry out the takeover threat. The state did not seem ready to incur considerable expense to get into the public power business, which it probably could not have performed any more efficiently than LILCO. (The same people except at the very top would probably have managed and operated the machinery.) Top LILCO management did not want to lose their jobs through a takeover and they could not afford the risk of bankruptcy, which the PSC through its rate-setting power could surely have forced. Thus LILCO dared not make good on its threat to operate Shoreham in the face of state opposition. Furthermore, LILCO's financial status was grim: a low rating for its bonds, $800 million in debt at interest rates between 13 and 17 percent, no rate increases for three years, and no access to capital markets for the past two years. A LILCO official said: "It was clear that the company would go down the tubes if it didn't settle."[30] LILCO may have had some less obvious reasons for abandoning Shoreham as well. A LILCO vice president said: "We want to usher in a new era of cooperation with government officials." *New York Times* reporter Matthew Wald wrote: "If they close the reactor people will stop bashing them over the head."[31]

In a further blow to LILCO's financial health, on December 5, 1988, a federal jury ruled that LILCO owed about $7.6 million in damages to Suffolk County under the Racketeer Influenced and Corrupt Organizations Act (RICO), for misleading state regulators about the cost and schedule of Shoreham construction in order to obtain rate increases in 1978 and 1984. The verdict was handed down four days after the death of the New York–LILCO settlement, prompting *New York Times* reporter John Rather, who observed the confusion in Albany, to write: "it has been the nature of the Shoreham story to top itself repeatedly. What appears briefly to be the denouement [sic] is superseded and cast in another light by what happens next."[32]

Since the federal racketeering law allows the recovery of treble damages, the judgment was tripled to $22.8 million. Damages could have soared to $1.8 billion or higher if federal district court judge Jack B. Weinstein were to uphold the verdict and allow all of LILCO's customers in Long Island and Queens to file a class action suit against the utility to collect refunds for having been overcharged. (Suffolk County, which had brought the suit two years before, had been seeking $2.9 billion in damages, an amount that under RICO could have been tripled to $8.7 billion.) Foes of LILCO were quick to take advantage of opportunities stemming from the jury's verdict. Irving Like said that a class action ruling could bankrupt the company, permitting LIPA to carry out a takeover. And the Suffolk County legislature lost no time in passing a "Sense of the Legislature Resolution" demanding that NRC stop consideration of an operating license for Shoreham until it could evaluate the safety and financial implications of LILCO's RICO liability.[33]

Judge Weinstein overturned the ruling on February 11, 1989, on the grounds that RICO, the basis of the Suffolk County suit, did not apply to a rate regulation case. The issue of electric rates, said the judge, properly belonged to the state PSC rather than to juries in a federal court. LILCO said it would attempt to negotiate a settlement rather than face the uncertainty of further appeals. The case was eventually settled, with the assistance of the PSC staff, when a court-appointed mediator reached an agreement with a lawyer representing LILCO's million ratepayers, whereby LILCO's proposed rate increases of 5 percent per year (as in the agreement with New York State) would be offset by $390 million in rate relief over the next decade. (LILCO would also pay $10 million in attorneys' fees.) The ratepayers agreed to forgo further court action, but Suffolk County refused to settle until a provision was included to close Shoreham.[34]

The governor revived his settlement offer to LILCO in March 1989, this time without the requirement for legislative approval. In assuming sole responsibility for closing the deal, the governor said with self-congratulatory eloquence: "And when we build that brass plaque on the beautiful seeded ground where there would have been a nuclear facility, where children are playing, gently swinging back and forth, let the plaque say no legislators' names."[35] As before, LILCO agreed to sell the plant to LIPA for $1, for the sole purpose of decommissioning the newly completed reactor. In return, LILCO would receive assistance from New York State in building new baseload capacity and about $2.5 billion in federal tax write-offs. The "financial regulatory asset" scheme of the first offer was abandoned. Rate increases for LILCO, a key sticking point in the original settlement, were not guaranteed; they were to be determined by the PSC and would be the commission's responsibility. LILCO agreed not to operate Shoreham even if a license was issued, so long as the deal was approved by the various state agencies (NYPA, LIPA, and PSC) by April 15, 1989. Once that condition was met, LIL-

CO would not operate Shoreham unless the agreement was disapproved by shareowners.

The agreement was silent about a pending lawsuit by LILCO against Suffolk County, the town of Brookhaven, and the Shoreham-Wading School District (where Shoreham was located) for overassessment of the Shoreham plant from 1984 onward except to urge LILCO to engage in "good faith" negotiations. The failure to include some kind of condition to protect the county and local governments from having to make refunds to LILCO emerged later on as a very serious flaw in the final settlement agreement.

The agreement came at a crucial time, within days of the issuance of an OL by the NRC. The PSC signified its approval, granting a temporary 5.4 percent rate increase to LILCO in February 1989, the first in three years, conditional on LILCO's not operating Shoreham. This provided the utility with an additional $97 million in revenue for the next year. The PSC followed this up with a financial recovery plan for LILCO, granting another 5 percent increase on December 1, and a third 5 percent increase on December 1, 1990. LILCO was to receive rate increases of up to 5 percent annually through 1998. Stating that its decision "represented a beneficial tradeoff between limited, short-term rate increases and greater uncertainty and litigation over the Shoreham issues," the PSC warned that if the settlement was rejected, Shoreham's ultimate fate would depend not only upon the NRC, "but also upon the courts, the results of LIPA takeover efforts, and indeed the experience of all other U.S. reactors." A single public service commissioner disagreed sharply, terming the "Shoreham deadlock as nothing more than a euphemism for the consequences of the political exploitation of anti-LILCO sentiment and the fear on the part of some uninformed people about the safety of nuclear power."[36] Investment bankers who were reported to have played a significant role in the negotiations and eventual settlement were said to be enormously pleased with the deal. Once the PSC agreed to increase electric rates, LILCO's credit rating was raised to just below investment grade.[37]

The settlement demonstrated, among other things, the importance of the PSC's rate-setting authority and the power of the financial community in both forcing and implementing a negotiated agreement. It also reflected a changed managerial climate at LILCO. Prior to the mid-1980s, the company's goal had been unequivocally the licensing of Shoreham. LILCO's behavior changed when Dr. Catacosinos—called by Fabian Palomino the "John Wayne of the nuclear industry"—became president in 1984 and, according to McDonnell, announced that his goal was to "save the company from bankruptcy. And everything about the company other than its existence was negotiable, including the licensing of the Shoreham plant. Clearly, he represented the shift from the old administration, which had made Shoreham the issue, and he made the existence of LILCO the issue."[38]

### Ending the Shoreham War

After Governor Cuomo's handpicked trustees took office, NYPA (which was to build new power plants for LILCO customers and decommission Shoreham) supported the Shoreham settlement by a three-to-two vote on April 12, 1989. LIPA approved the settlement the next day, and LILCO's board of directors did so a few days later. Soon after, the IRS approved a $2.5 billion tax write-off, which would enable LILCO to claim a substantial tax credit each year for an estimated seven years, one of the key conditions in the agreement. On June 29, 1989, LILCO shareholders gave overwhelming final approval to the utility's plan to abandon Shoreham.

Despite these positive developments for LILCO, not everyone was pleased with the broader implications of dumping Shoreham. The nuclear industry vigorously opposed it. "They felt that Shoreham was victimized . . . and that the settlement should be resisted and prevented in every way." Freilicher thought this position was misguided and shortsighted: "In addition to all of the other problems of the nuclear industry, if they were going to make it hard for a company to get out of a project, they would have made it more difficult for industry to enter into projects in the first place."[39]

Secretary of Energy Watkins characterized the Shoreham deal as "stuff and nonsense," if not "one of the [most] foolish deals in the nation's history," and dismissed the dismantling of Shoreham as "utterly irresponsible." He vowed that the Bush administration would continue its efforts to save the plant from "senseless destruction." The deputy secretary for Energy, W. Henson Moore, was less temperate, stating: "People shouldn't be making decisions on the basis of political rhetoric or hysteria"; he urged the NRC to prohibit LILCO from taking action to close the reactor. The *Wall Street Journal* wrote that "Governor Cuomo embraced a position that is simply anathema to the best interests of this country's energy and environmental needs." And the *New York Times* warned: "New York is about to commit a blunder so monumental that, like the pyramids, it may prompt future generations to marvel at the ruinous excesses of human folly." In addition to Governor Cuomo and the NRC, the *Times* blamed Long Islanders who "punished any politician who dared tell them the truth. . . . Now they face the worst of three worlds: higher price electricity and blackouts and heavy reliance on foreign oil."[40]

## A Bailout for LILCO

If LILCO's opponents had hoped for a victory that would punish the utility they had learned to hate, they were badly disappointed. The settlement was extremely beneficial for LILCO. It removed the threat of bankruptcy, renewed the utility's access to capital markets, and cleared the way for rate increases for the next ten years. LILCO was probably better off than it would have been if it had operated the plant: it faced no financial risks as a nuclear plant owner, enjoyed greater rate certainty, and achieved peace for itself. The utility was able to resume payments for owners of preferred and common stock dividends in October 1989, the first such payments since 1984, and to pay its top executives more than $250,000 in bonuses, retroactive to 1988, for their work on the Shoreham settlement. President Catacosinos received a bonus of $149,000 and a raise of $25,000, bringing his salary to about $350,000 per year. (It was also reported that Dr. Catacosinos had been given a ten-year retirement package worth $5 million.) A close observer noted that LILCO was "enormously skillful in its negotiations":

It put together a group of very talented experienced people in a highly coordinated effort. . . . It confronted many different uncoordinated groups including the media, the governor, the legislative bodies in Albany and Congress, the feds, financial institutions, etc. . . . What happened is that LILCO held off, broke through on the NRC front and forced a settlement which was ultimately fed by the blood of Shoreham. . . . It meant that the governor had to kill the project on LILCO's terms.[41]

Suffolk Congressman Downey agreed that LILCO was better off with the rate settlement than with the plant. "Had there been an accident or misstep, people would have shot the heads of the utility. . . . [There would have been] large civil unrest on Long Island . . . and you would have had the governor saying: 'I agree with these people.' A utility can't operate and flaunt the governor and all local officials and say 'screw you.'. . . If that plant had operated and had had one problem, it would have never gone on-line again."[42]

A lot of people were outraged, but they were relieved as well. Irving Like reflected a more general view when he called it "a bittersweet situation. Everybody is happy that the Shoreham situation is resolved but everybody is unhappy that it's such a sweetheart deal for LILCO."[43] Governor Cuomo, in particular, came in for his share of criticism. In achieving his primary goal of shutting down Shoreham, he had agreed to a financial settlement that placed the major burden on the ratepayers. A New York business leader commented: "I can't imagine why any responsible political office holder would sign an agreement like that." A former PSC member noted: "It's a lousy settlement because . . . no one speaks for the consumer." A LILCO official said: "The unfair thing was that the public was led to believe initially, and the politicians stated this, that if indeed the plant was abandoned, the public wouldn't pay for it. How could they pay for something they didn't agree with?" And a former LILCO vice president agreed: the "losers were consumers unless you accepted the notion that it's cheaper not to operate the plant."[44]

But was it cheaper not to operate the plant? New York officials maintained that electric rates would have been higher, at least for a decade, had Shoreham operated. The New York CPB explained:

First, the Settlement will permit LILCO to operate with lower capital costs and lower interest coverages than if Shoreham were in service. Eliminating Shoreham will substantially reduce the risk of investing in LILCO. The risk will be reduced further by having a long-term rate plan subscribed to by the utility and . . . the Commission. This moderation of LILCO's risks will reduce the cost of money to the Company and ultimately the cost of service to ratepayers. Second, studies purporting to show that Shoreham would produce cost savings failed to account adequately for the substantial capital additions that would have been needed to meet changing NRC requirements.[45]

A specialist in nuclear power economics who testified for the CPB agreed with the would-be cost savings. His update on March 22, 1989, of a 1985 CPB study on the economics of closing versus operating

Shoreham concluded that running Shoreham (in 1989 present value) would cost LILCO ratepayers at least $900 million more, over a thirty-year period, than closing the plant. (This figure was substantially more than the finding of the earlier study; that is, that running Shoreham would cost ratepayers at least $300 million more over thirty years than if the plant were not operating.) His conclusion in 1989 was based on changes in key costs affecting the economic results, including "replacement energy costs" (that is, declining oil and gas prices), "replacement capacity costs" (combustion turbines, combined cycle units, and coal units as compared with nuclear), "Shoreham operation and maintenance costs" (higher for nuclear units than for fossil plants), and the cost of "Shoreham capital additions" (higher for nuclear than for fossil).[46]

The PSC staff believed that operating Shoreham would "entail greater rate increases (at least for a decade), as well as the financial risks of operating a nuclear plant, and the certain opposition of many other parties." The Public Service Commission itself took a more conservative view, stating that "rate increases are inevitable on Long Island, whether or not Shoreham runs, whether LILCO goes bankrupt, and whether there is a LIPA takeover. Predictions as to which alternative will be least expensive for ratepayers depend heavily on many future unknowns; but based on our evaluation, the settlement benefits ratepayers."[47] The PSC argued too that rejection of the settlement would lead to continued public opposition to Shoreham, resulting in divisive wrenching of the Long Island community and further litigation: "[This] . . . would leave plans to meet Long Island's capacity needs in a quandary, sap management attention, drive up expenses, and continue to muddle any equation of the utility's revenue requirement, making establishment of just and reasonable rates extraordinarily difficult, if not impossible."[48]

Both the chairman of the PSC, Peter Bradford, and the director of the CPB, Richard Kessel, were less equivocal when they spoke for public consumption, stating: "Long Islanders would pay $2.5 billion more with Shoreham than without it by the year 2002" because of the "higher financing costs and post startup capital requirements of nuclear plants." (The basis for these assertions is uncertain.) Matthew Wald probably had the correct notion when he pointed out that "answering the questions [of winners and losers] requires a room full of financial

and energy analysts, a computer and a crystal ball." The imponder-
ables would depend, among other things, on how much the oil that
would have been saved would have cost, how reliable the plant would
have been, and how many expensive new regulations would have been
imposed by the NRC.[49]

Last, but by no means least, in evaluating the settlement it is important to note that the agreement with the state embodied the most
sought-after and most valued gift of all for Long Islanders: the belief
that they were safer as a result of Shoreham not operating. The head
of the CPB said: "The bottom line is that the plant is closed. . . . How
do you measure the cost in terms of human life?"[50] Three years later,
the PSC chairman noted that Shoreham turned out to be an "economically sensible solution":

> What was done largely for one set of reasons [that is, safety] has in fact worked
> out well in economic terms. . . . Nothing that's happened has given New York
> any cause to regret it so far. Interest rates are actually lower than we had fore-
> cast, and demand is lower than we had forecast, and oil prices are lower than
> we had forecast. If the plant had run, there is no doubt that electricity prices
> would be higher.

Bradford also pointed out that Long Island was much less vulnerable to Middle East politics than even the PSC had thought, because
they had converted a lot of their oil-fired plants to burn oil and gas.
When oil prices spiked, they burned natural gas.[51]

## Mario Is the Single Person to Credit

It is generally agreed that Governor Cuomo's opposition and his
ability to compel a settlement were the most important actions that
killed Shoreham. As Congressman Downey said: "He had both the
courage of his convictions and the power of his office."[52] Once the governor came out against the plant, that was the end. Called by many the
champion of states' rights, Governor Cuomo consistently questioned
federal licensing of the Shoreham plant. It was he who commissioned
the Marburger panel in 1983 to examine Shoreham's economics and
safety, challenged the licensing of Shoreham before the NRC, appointed new PSC members and members of NYPA who shared his views
about closing the plant, and ultimately took responsibility for reaching
a settlement with LILCO that was unpopular with many of his constituents, including members of his own political party.

At the same time, his sincerity in opposing the plant is questioned by many in the federal establishment and elsewhere. It is claimed that Cuomo, like the Suffolk politicians, responded to what he thought was a popular sentiment against nuclear power on Long Island, that he saw his opposition as a "politically desirable act" and "exhibited great skill in determining which way the wind was blowing." An NRC attorney, Bernard Bordenick, reflected the general agency view when he said:

Politicians cater to what their constituents want. That's how they get elected. And Mario Cuomo didn't say, "I'm a politician." . . . He said: "You can't evacuate Long Island." And that's nonsense. . . . People would have demanded that Shoreham go into operation if they had put on their light switches and nothing happened because there wasn't enough electric power on Long Island. . . . He didn't lay all the facts out to the people. He was intellectually dishonest. . . . I would have liked him to be up front about his decision and who was going to pay for it.[53]

LILCO's attorney was critical of the arguments that he remembers the governor using, saying:

If you don't want a nuclear power plant built . . . [because] you just generally can't stand that technology or utility bureaucrats, you are rarely going to cast your opposition in those terms because they would be losers. What you say, as Governor Cuomo did when he decided for political purposes largely, to oppose the plant: "Do you want to harm pregnant women and unborn children? . . . It's going to irradiate you in your sleep. . . . You'll be trapped like rats if it has an accident." You say those sorts of things. They have a better press, they elicit a much greater response among people who have no real handle on the facts, and they might actually win.[54]

Governor Cuomo's supporters point out, however, that it was not a popular role at the time he took it, that he might well have shifted the blame. According to Kessel, Cuomo could have fought LILCO and not arrive at a settlement. "He would have lost. NRC would license. They would be the villain. . . . But that's not Mario Cuomo. Took an extraordinary amount of courage. He knew what the result would be and that people would be unhappy with the rates."[55] Governor Cuomo had little to gain, said his deputy secretary and energy advisor, Frank Murray:

When the governor first announced his opposition to licensing the plant, in terms of public opinion, he was in the minority. People like to look back and think that the governor was doing this to capitalize on the political wave in the eastern end of Long Island, and people characterized this as pandering to the opinion polls. Quite the contrary. The polls were the other way around. When the governor was opposing the plant early on, you didn't have more than a small group of activists. . . . As governor, he was deeply concerned about the adequacy and reliability of the electrical system. He didn't want the lights to go out in Long Island, particularly in a hot summer afternoon. That's not a political formula for success. He was convinced that there were alternatives available to meet the supply needs, not only then but projecting out for the next ten or fifteen years, all without Shoreham.[56]

Surprisingly, LILCO's former president agrees with the substance of the latter remarks, stating that he was convinced Cuomo was "an honest governor" who believed in what he said. LILCO's Freilicher was similarly convinced. Political analyst Ben J. Wattenberg, a Shoreham supporter, castigated Governor Cuomo for killing Shoreham, but wrote that a talk with the governor had persuaded him Cuomo was acting out of "conviction, not hot seat political expedience."[57]

Those who believe in Cuomo's sincerity stress that he was not exhibiting some kind of blind opposition; the governor was not opposed to nuclear plants per se. "He's skeptical about it," said a high-ranking state official: "He questions the economics . . . he questions the safety . . . he questions whether there's a solution to the waste problem. On the other hand, he's an intelligent enough man to know that science may yet find a way of solving some of these problems." As for his primary motivation: "the governor was convinced, and remains convinced to this day, that the plant was unsafe, . . . that no evacuation was possible, that there was no place for people to go." Like said that the governor seemed "almost constantly distressed at that particular point [that is, the evacuation issue]; that in the case of Shoreham it was an unacceptable situation. It had to do with the geography on Long Island."[58]

Less charitable observers of Cuomo, such as the *Wall Street Journal*, characterize him as "the nuke killer" and argue that "the so called 'safety' issue was a red herring from the beginning. There hasn't been a nuclear-radiation fatality in the United States in thirty years." The governor's concern about evacuation from Long Island was similarly mocked as "nonsense because they don't have to evacuate the whole of Long Island. It's like saying you can't have Indian Point because you can't evacuate New York City." In countless editorials, the *New York*

*Times* criticized the governor's "political timidity" and faulted him for not taking a "statesman-like position" to educate voters, calm their fears, and make Shoreham the safest nuclear plant in the country. And the Scientists and Engineers for Secure Energy, Inc., a pro-Shoreham group, criticized the governor for failing to consult with them, stating that his decisions were made with "less than informed knowledge."[59]

Whatever the motivating factors, it is clear that the governor's position carried great weight. LILCO's chief nuclear counsel, Taylor Reveley, said:

Up until the point that the governor began to actively, directly and indirectly, overtly and covertly, intervene against the plant in a big way, opposition to it was in a very small part of the community. And across the rest of the community, it was manifested in a growing distaste for LILCO and a growing skepticism. But the opposition was very thin. Sort of thing that could easily have been motored on through. Until the governor, who had enormous credibility and enormous political standing, began to say that *he* believed the plant was not safe, and *he* believed that you couldn't have effective emergency planning, and *he* believed that the federal nuclear establishment was at best a bunch of bozos in whom you could repose no confidence. "So NRC says it's safe. So what?"[60]

In fact, back in Washington, the NRC was indeed finding that the plant was safe. Despite Governor Cuomo's activities in New York, despite the settlement and the confirming acts of other governmental agencies, the NRC and the administration were still actively engaged in discovering ways to authorize the plant to operate. The future was still uncertain. "These are unprecedented situations," said Suffolk County's attorney: "The one thing . . . we can say for certain is that nobody knows what would happen in the absence of this agreement. . . . The other thing we know . . . for sure is that the NRC would like to give Shoreham a license."[61]

THE HANGING GARDENS OF SHOREHAM

*Drawing by W. Miller; © 1989 The New Yorker Magazine, Inc.*

# 7 Dead on Arrival

## NRC Licenses a Canceled Power Plant

Various federal agencies including the NRC made a final futile effort to enable the Shoreham plant to operate. After lengthy proceedings, NRC found that emergency planning at Shoreham was adequate and issued a full-power license. The license, however, had no impact; one week before, LILCO's board of directors had agreed to sell the
plant for $1 to LIPA, which planned to dismantle it. Despite DOE's per-
sistent attempts to save the plant, decommissioning was finally accomplished. Yet the plant has left Long Island a continuing legacy, the
highest electric rates in the nation.

### It's Alice in LILCO Land

While Governor Cuomo in New York was making plans to close
Shoreham, the federal government was equally determined that Shoreham would operate. President Reagan entered the Shoreham fray in
November 1988, issuing an executive order that gave FEMA broad new
authority to draft and implement evacuation plans for nuclear power
stations whenever state or local governments refused to do so.[1] Although the order contravened an earlier Reagan promise not to impose
federal authority over the objections of state and local government to
an emergency evacuation plan for Shoreham, it did remove the major
obstacle to the licensing of both the Shoreham plant and the Seabrook

plant in New Hampshire. White House officials were reported to view the order "as the opening salvo in a battle to return the nuclear power industry to robust health." Joseph McDonnell, a LILCO vice president, thought it might make an important difference in the troubled campaign to get Shoreham licensed. "With the federal government participation not only in the event of a national emergency but also in the planning for that possibility," he said, "I can't see any reason why these plants [Shoreham and Seabrook] should be stopped from getting a license." The *Washington Post* more realistically called the executive order merely "a symbolic gesture that . . . isn't going to do much to strengthen local communities' trust in federal nuclear policy or the people who carry it out."[2]

The DOE took measures to challenge developments in New York State. In a "prohibited ex parte" communication, Secretary of Energy John S. Herrington urged the NRC "to move expeditiously to complete the decisionmaking on whether to issue a full power license for the Shoreham nuclear power plant. . . . The enormous burden on our nation's energy security, the ratepayer and the federal taxpayer argues for a swift and final decision."[3] NRC did not need DOE's prodding to counter Governor Cuomo's efforts to negotiate a shutdown of Shoreham. Responding with a narrow legalistic interpretation that the pending agreement between New York and LILCO did not constitute "a formal request for withdrawal of the license application," NRC informed Congressman Hochbrueckner that "regulatory agencies such as [itself] . . . have an obligation to continue licensing proceedings."[4]

Notwithstanding the New York–LILCO agreement not to operate the plant, the NRC asked FEMA to participate in a three-day "full participation" exercise of LILCO's emergency plan, which would, it was hoped, correct the "fundamental flaws" found in the earlier drill. FEMA was hesitant about spending money for an apparently meaningless exercise but ultimately went along with the request. The exercise was labor-intensive and expensive, involving about 160 federal employees and 3,000–4,000 LILCO workers at a cost of about $2 million. This time around, FEMA found "adequate overall preparedness" by LILCO personnel and "reasonable assurance" that public health and safety would be adequately protected in the event of an accident at Shoreham.[5]

At about the same time, a three-member Atomic Safety and Licensing Board named for James P. Gleason, the chairman, asked New York State and Suffolk County (and the town of Southampton, which had joined the other intervenors) to prepare their own emergency evacuation plans for Shoreham or to explain what action they would take in the event of a radiological emergency at Shoreham. The alternative, warned the board, would be a dismissal of their objections to the LILCO evacuation plan. New York and Suffolk at first failed to comply, stating that they could not "speculate" about their actions or their resources in an emergency. Belatedly, the two governments submitted a previously undisclosed "Emergency Operation Plan" that had been prepared many years before, indicating to the board that they knew more about emergency plans and resources than they had previously been willing to divulge. Incensed, the Gleason Board charged them with "a sustained and willful strategy of disobedience and disrespect for the Commission's adjudicatory processes" and ordered them dismissed from *all* commission proceedings (not just the one this particular board was conducting) on the Shoreham license.[6]

The board had legitimate reason for its righteous anger over the withholding of important information on the emergency planning issue, a tactic that had dragged out an already burdensome proceeding and wasted everyone's time. Dismissal would have seemed no more than the state and the county's just deserts for such obstructive conduct, had this been an ordinary adjudication (that is, one in which no one except the parties themselves had any significant stake). But the Shoreham OL proceeding was obviously infused with matters of great public interest. A sound decision on the emergency planning issue could be of vital importance for public health and safety, and the state and county governments would have to play the central role in any real emergency. Throwing them out of the decision-making process undoubtedly felt good to the harassed licensing board, but the action revealed the adjudicators' shortsighted perspective. One board member dissented from the dismissal order, noting: "it seems to me unwise to reject the present governments' participation in emergency planning, an area where the Commission's rules have traditionally given great deference to local expertise."[7]

An NRC appeals board reversed the part of the board's order that dismissed the state and county governments from the entire licensing case, on the grounds that the licensing board had overstepped its authority.[8] The appeals board left standing the dismissal of the state and county from the particular adjudication over the "realism" of LILCO's emergency plan. Thus it became a settled finding that, in the event of a real emergency at Shoreham, the state and county governments would do their very best to follow LILCO's plan, even though their representatives had vowed they would do no such thing. With the adequacy of emergency planning now established in law (if not in fact), at least before the licensing boards, both the commission staff and another licensing board recommended in response to a long-standing LILCO request that LILCO be allowed to operate the plant at 25 percent power. Observing the performance of the multiple licensing boards, the New York governor's energy advisor remarked on the "clear perception that their mission was not to evaluate the merit of off-site emergency planning but to find a way to license the plant."[9]

## No Less Arrogant than LILCO

The formal end came for the intervenor governments on March 3, 1989, when the commission exerted its plenary authority and dismissed the governments from all Shoreham licensing proceedings as a sanction for obstructive conduct. Acknowledging that federal-state relations should be marked by "mutual cooperation rather than confrontation," the commission nevertheless piously declared that "an unwillingness ever to punish misconduct would be an abdication of responsibilities."[10] In a move that was perceived by the Shoreham opposition as extraordinarily high-handed, the commission thereupon eliminated from the entire Shoreham proceeding the parties with the most expertise concerning emergency evacuation concerns, the main legal issue in the case. The license was now formally unopposed.

New York thought it was "outrageous" to throw the state out of the proceeding. Although they understood that the NRC was annoyed beyond measure and believed the state was just being obstructive, New York officials nevertheless faulted the NRC action. "Even if that perception was an accurate portrayal of what the state was doing (which I don't believe)," explained Murray, "the perception that you don't want to hear criticism about what you're doing feeds the perception that you have a defined objective in mind that, come hell or high water, you're going to get. And if you have to throw the State of New York out . . . to get there, so be it." Irving Like agreed. "Even if you could find some justification for eliminating the state and county," he said, it was "an unwise act. It was perceived as an arbitrary, high-handed act on the part of somebody who was bent on issuing a license." Peter Bradford thought the state and county could have successfully appealed their expulsion had they chosen to seek immediate judicial review of the commission's action: "Whether it would have resulted in a stay of the operating license or not is an open question."[11]

Commission staff viewed the expulsion differently. They took advantage of the lack of formal opposition to move quickly on emergency planning at the Shoreham site, finding that Long Island's geography, highway system, and other features did not make emergency planning fundamentally more difficult than elsewhere.[12] The commission agreed that Shoreham was a safe plant for operation and issued a full-power operating license. Shoreham thus became the first U.S. nuclear plant to be licensed with a utility-developed emergency plan rather than one formulated by state and local governments. Some observers looked upon the commission vote as a "face-saving" gesture intended to assure nuclear power advocates that the federal government would not allow a completed plant to be blocked because of state and local opposition. By giving the plant a license, LILCO said, "the NRC affirms the principle that it is the Federal Government—not states and counties—that rules on the safety of nuclear power plants."[13]

The operating license was issued on April 20, 1989, twenty-two years after the utility originally announced its plans for a nuclear reactor, and four years after construction was completed.[14] It came a little more than one month after LILCO and New York State had entered into the settlement agreement that assured Shoreham would never operate, and less than a week after LILCO's board of directors had approved the settlement. NRC's issuance of a full-power license for Shoreham amounted to a certificate of good health for a patient who has just died. The *New York Daily News* likened it to "throwing a bon voyage party for the *Titanic* six weeks after it hit the iceberg."[15] NRC's finding that the plant was safe was similarly irrelevant. Proponents of the plant were convinced from the beginning that it was safe and did not need the NRC to tell them so. The opposition regarded NRC's finding as a foregone conclusion, and it confirmed their impression of NRC as an advocate.

## NRC Essentially Inconsequential

When NRC chairman Lando W. Zech Jr. was questioned about the significance of a decision to license a plant that was about to be abandoned, he articulated a narrow (but legally accurate) vision of his official role, saying: "As Chairman of the regulatory commission, that's not my responsibility. As an American citizen, I would call it a waste."[16] There is little argument about the waste. It was estimated that, since 1983, LILCO and its opponents had spent more than $75 million in the licensing battle, with taxes backing the effort to shut the plant and electric bills funding the effort to open it. It was an extremely lucrative litigation for the half dozen or so law firms hired by LILCO and Suffolk and for their consultants and lobbyists. Moreover, in the thirteen years or so that LILCO's application for a license had been before the NRC, LILCO had paid the U.S. Treasury more than $3 million to reimburse NRC for costs tied to the proceedings. Expenses rose still higher when the several hundred hearing days, one thousand exhibits, testimony from over seven hundred witnesses, and the more than sixty-five thousand pages of hearing transcripts are included.[17]

Although Governor Cuomo called the Shoreham license "an emblem of hypocrisy," DOE Secretary Watkins praised it as "perhaps the most important development for the nuclear industry in the U.S. in the past decade. . . . Because Shoreham is the right energy source at the right time for the right place, its operation should stand as a symbol of the promise and potential of nuclear power in America." DOE's deputy secretary warned: "We intend to throw up every roadblock [to its closing] we can. And if we have to create some, we'll do so." Watkins was more outspoken: "If activists can prevent things from being built, by God, I can prevent things from being shut down when it is stupid."[18]

Watkins turned out to be wrong about this, but the proponents of Shoreham did not give up easily. In Congress, Representative Don Ritter (R-Pa) introduced an amendment to NRC's authorization bill that would bar the commission from transferring Shoreham's license to LIPA. (Under federal law, the NRC had to give its permission before LILCO could transfer Shoreham to different ownership and control.) Governor Cuomo, appearing before a House Interior Committee hearing, asked Congress to keep hands off. NRC was silent about the congressional move, thereby evoking criticism that it was sacrificing its independence by yielding too readily to congressional interference with its regulatory mission.[19] The Ritter amendment was eventually withdrawn at the request of Long Island representatives and never made it to the floor of the House for a vote.

Responding to remarks by White House chief of staff John Sununu that the Shoreham plant was a "national asset," Governor Cuomo argued that the ultimate fate of the plant was a matter of "states' rights"—that the state had "exclusive jurisdiction over the decision to operate a particular power plant." He urged President Bush not to interfere. In a departure from its usual practice, LILCO declined to comment about the differing views; now that its financial future was assured, LILCO was content to let the debate over decommissioning be strictly a governmental matter, "a public policy decision . . . that should be decided by the state and the federal government."[20]

## Mothballing, a Minimum Fallback Position

Governor Cuomo's position was not convincing to Shoreham backers or the administration. Insisting that Shoreham's dismantling was a mistake, plant proponents sought to "mothball" the plant, to put it in storage in case New Yorkers changed their minds and needed Shoreham to produce electricity in the future. "The ideal situation," as one business executive stated it, "was for the New York Power Authority to take it over and . . . suspend it. And that would have given time to check it all out and to come back and say to the public 'it's a modern safe plant,' and start it up the next time we have a Middle Eastern embargo of some sort when Long Island would have been desperate for power at any price."[21]

The DOE's object was to "delay" the dismantling and transfer to LIPA for several years, in case electricity was in short supply during the summer months. This left New York State and Shoreham opponents in the unusual position of pushing for speedy administrative action whereas the supporters were now arguing for lengthy delays. In addition to DOE and the President's Council on Environmental Quality (CEQ), plant supporters included Scientists and Engineers for Secure Energy (SE2), a group of about two thousand scientists, who thought it was their duty "to tell people they are doing something that they don't understand," and the Shoreham–Wading River Central School

District (School District), which derived 90 percent of its property tax revenue (about $28 million annually) from the plant, making it one of the richest districts in the country. They petitioned NRC to halt dismantling until an Environmental Impact Statement (EIS) could be prepared that would consider a full range of alternatives, including plant operation. (Environmental groups were quick to point out that this concern for "environmental impacts" represented a new posture for the Energy Department and the others.) Such an environmental review would certainly delay the state's taking control of the plant, adding two or three years to the shutdown process.[22]

Mothballing—or keeping the plant intact so that it might produce electricity if conditions changed—was estimated to be an expensive option. A LILCO vice president pointed out that maintaining Shoreham would cost $75 million a year without adding in the cost of NRC's changing requirements. And the head of New York's CPB estimated that mothballing for the years 1994–2002 would increase LILCO's need for revenue by about $2 billion, a cost that would be passed on to ratepayers. The higher cost would presumably come from the number of employees to be kept on to maintain the plant in operating condition. The PSC chairman warned that the commission would not approve such a payment by LILCO customers, stating: "If DOE wishes to advocate mothballing, they need also to be prepared to pay the bill for it."[23]

DOE asked NYPA to withdraw from the agreement with LIPA on the basis that the contract was illegal and inconsistent with the "public interest." Richard Flynn, chairman of NYPA, responded that he held a different view of the public interest. Noting that a substantial portion of the Long Island population and their elected representatives had voiced opposition to Shoreham, Flynn wrote: "The duties of representing the public interest in a democracy require elected officials, in New York and elsewhere, to listen to the public's statement of its interest and to act on troublesome and difficult-to-resolve issues."[24]

DOE was persistent, soliciting the Justice Department to join in the effort to challenge the agreement transferring Shoreham to LIPA. In a formal letter urging the NRC to delay action, the secretary of energy reiterated that "Shoreham's destruction would be a colossal mistake, . . . contrary to every principle associated with the establishment and maintenance of a sensible national energy policy." In September 1989, DOE became a plaintiff in a lawsuit brought by the Atlantic Legal Foundation (a right-wing organization that defended groups against "excessive governmental regulation") to gain an injunction against the implementation of the New York–LILCO agreement. The case was rejected by state supreme court justice F. Warren Travers who passed the case on to a higher court, the appellate division of the New York State Supreme Court.[25]

Mothballing was not seriously considered. The NRC rejected the petitions to consider alternatives to dismantling on October 17, 1990, stimulating criticism from one group of scientists that it "prematurely aided the decommissioning of the plant. . . . It should have put up procedural roadblocks."[26] The clear implication here was that the NRC was rushing with unseemly haste to dismantle a plant that it had taken much too long to license.

## Decommissioning the Plant

Decommissioning would be another first for Shoreham. Eventually nuclear plants all over the United States, many of them bigger than Shoreham and far more radioactively contaminated, will have to be decommissioned. But it had not happened yet. There were no precedents for demolishing a plant of this size. It was not certain how long it would take or what it might cost to dispose of the steel and concrete made radioactive by Shoreham's brief period of low-power testing. New York and Suffolk had opposed low-power testing on precisely the grounds that radioactive contamination of the plant would be an expensive impact with no benefit if full-power operation never occurred. The commission had brushed that objection aside.[27] Now the scenario the commission had scoffed at had come to pass. A reporter who visited Shoreham described the potential magnitude of the decontamination job as one of dismantling "18,000 tons of steel and 130,000 cubic yards of concrete, enough to pave a highway for about 25 miles, . . . 2,200 miles of electric cable, a spaghetti-like tangle of piping and conduits, 20,000 hand-operated valves and 2,000 motor operated valves, some of them as large as a small car."[28]

Decommissioning was a tortuous and complicated process, involving NRC review of every modification, no matter how trivial. Before ownership could be transferred to LIPA, LILCO would have to file for and obtain a "possession only" (POL) license from the commission. This meant the company would be allowed to "possess" the plant but not operate it, a status involving reduced—but by no means negligible—maintenance. Then LIPA, an organization with few financial resources, virtually no staff, and no nuclear experience, would have to obtain a license from the NRC establishing its technical and financial ability to assume ownership of the plant. LIPA would also have to file a decommissioning plan with the NRC and apply for permission to start dismantling the reactor. Then LIPA and NYPA would seek an amendment to the reactor license to move the radioactive fuel out of the reactor building and store it on-site. Eventually, the reactor fuel would have to be transported to another location. Had the plant not been made operational at 5 percent power, it would have needed simple demolition, not decommissioning. The whole process would have been far simpler and cheaper.[29]

LILCO took the first steps toward dismantling in the summer of 1989, moving uranium fuel rods from the reactor to an adjacent spent fuel pool and reducing its staff by about one-fourth. NRC's acquiescence in LILCO's request to disassemble its emergency planning organization and reduce its property damage insurance coverage in 1990 appeared to signify a clear recognition by the commission of the plant's "defueled" status and its commitment to abandon operations.[30] Then Iraq's invasion of Kuwait in 1990 provided nuclear power advocates with one final opportunity to revive Shoreham.

As the price of oil and gas rose, so did the shrillness of the voices demanding that Shoreham be kept intact. The *New York Post* thought that "Shoreham can still make a difference" in safeguarding the nation's oil supply. The *New York Times* claimed that operating Shoreham would save LILCO twenty thousand barrels of oil a day. The *Wall Street Journal*, not for the first time, took the opportunity to criticize state regulators who interfered with utility investment in nuclear energy, with Shoreham being the most "egregious example." The *Christian Science Monitor* thought the United States could use about 150 major new oil-saving nuclear plants this decade.[31]

In a renewed effort of White House involvement, the CEQ joined DOE in demanding a review of the environmental consequences of decommissioning an operational nuclear plant. DOE Secretary Watkins wrote to the NRC chairman repeating DOE's request to avoid action to hasten Shoreham's demise during the domestic energy crisis. In a second communication, a twenty-five-page "Amicus Submission," he threatened to use authority provided him under the DOE organization act to force the operation of the plant in the event of an "energy emergency." The New York PSC chairman responded in kind, calling DOE's comments "energy crisis creation."[32] DOE did not follow through on its threat.

**A Watershed Decision**

NRC seems to have walked something of a fine line, akin to a tightrope, in responding to the contradictory messages. Although the commission did not move as quickly as New York would have liked (speed would have been helpful in saving some money for Long Island consumers), NRC did eventually reject the demands of DOE and the other intervenors who "wanted to turn a license to operate into a sentence to do so."[33] Trying to placate all sides, NRC advised its staff to take a "middle ground" between those who were advocating speedy dismantling in preparation for complete decommissioning (that is, LILCO, LIPA), and those (like DOE) who were seeking to maintain the plant in full operating condition. But NRC could not please everyone, and its subsequent actions seemed to indicate that it was fed up with the Shoreham debacle and wished to rid itself of participation in that plant's business as quickly as it could.

In October 1990, NRC rejected a petition from SE2 and School District (and a request from CEQ) claiming that federal law required the commission to prepare an impact statement before it granted LILCO a POL license for the nuclear materials on-site. The statement would have had to include an evaluation of "resumed operations" as an alternative to decommissioning, an option that the NRC found to be of "speculative feasibility," that would require a significant change in government policy. The commission found further that LILCO was legally entitled to make "an irrevocable decision" to abandon Shoreham; to consider alternatives to this decision was "beyond Commission consideration." (What the commission would review, it said, was "the method" used for decommissioning but not the decommissioning decision itself.)[34]

In point of fact, the commission could have discussed the relative costs and benefits of operating the plant, as an alternative to destroying it, even if the commission could not order that the plant be operated. NRC staff thought that one could argue the NEPA obligation either way but that, if the commission addressed whether it might be better to operate Shoreham than to tear it down, the commission would get involved in complicated energy policy questions. The simplest tack to follow was to take the position that NEPA did not apply. The argument the commission gave was that NRC had no authority to order LILCO or anyone else to operate Shoreham, and therefore NRC was not obliged to look at that alternative.[35] If NRC had addressed the Shoreham operation alternative, the agency would probably have had to find that operating (or at least preserving) Shoreham was preferable to destroying it, and that would have gotten them embroiled in a conflict with LILCO, an outcome they wished to avoid.

LIPA called this "the watershed decision," but plant backers termed it an "abdication of the NRC's responsibility" to protect the environment. DOE, CEQ, SE2, and School District sought a "cease and desist" order of the defueling and destaffing action. This was denied by commission staff on December 20, 1990. Before the end of the year, LIPA submitted a Decommissioning Plan, an Environmental Supplement to Decommissioning, and an application prepared by Bechtel Power Corporation, the principal contractor for the Shoreham decommissioning, to transfer the license from LILCO to LIPA.[36]

The method proposed for dismantling the Shoreham reactor was DECON (short for "decontamination"), in which the equipment, structures, and radioactively contaminated portions of the facility are promptly removed or decontaminated to a level that permits the site to be released for unrestricted use. The alternatives to DECON are SAFSTOR (in which the facility is safely maintained for an interim period and then decontaminated) and ENTOMB (in which radioactive contaminants are encased in a "structurally competent" material such as concrete and placed under continuous watch until the radioactivity decays).[37]

## A "Test Case" for Nuclear Power

By the beginning of 1991, steps toward the decommissioning of Shoreham were well underway.[38] Fuel had been removed from the reactor core, staff had been reduced by two-thirds, security was cut back, and about forty different operating systems had been mothballed. Further dismantling would require the amendment of the Shoreham operating license to a POL license. NRC authorized a POL amendment of LILCO's operating license on June 12, 1991. NRC's actions elicited a blast from Energy Secretary Watkins who expressed strong disappointment. "It is a fitting irony," Watkins said, "that the NRC has basically given the last rites to Shoreham today since it is the NRC which has repeatedly affirmed the safety of the plant and the effectiveness of the evacuation plans."[39]

Issuing the POL was termed the "death blow" to Shoreham in that it indicated a "formal recognition" by the NRC that the plant would not be used to generate electricity. The decision allowed LILCO to dramatically curtail maintenance at Shoreham, allowing facility degradations that would make it virtually impossible to restore the plant. According to *Newsday*, this would save the utility up to $40 million over the next eighteen months. The decision denied another request by SE2 and School District to delay action pending the outcome of litigation challenging the settlement agreement.[40]

Backers of the plant, joined by the Justice Department, made a last-ditch effort to block the planned demolition and force an environmental review of alternatives. In lieu of a full EIS, Justice said that an "environmental assessment" would do. Both the U.S. Court of Appeals for the District of Columbia and the U.S. Supreme Court refused to stay the commission's issuance of the POL. The persistent petitioners and the Justice Department tried once more, in September 1991, to challenge Governor Cuomo's 1989 settlement agreement before the New York Court of Appeals. This too was unsuccessful.[41]

This ended the challenge by DOE to make Shoreham "a test case" of the administration's energy policy. DOE had not only failed in its long-standing effort to delay Shoreham's dismantling; it had also gained an unsavory reputation among those it was seeking to influence. Indeed, its effort may have backfired. As Peter Bradford said: "All the Energy Department's behavior does is reinforce the public perception of nuclear power as a highly controversial technology that needs to be shoved down the throat of a resisting public."[42]

On February 26, 1992, the NRC approved the transfer of the Shoreham license from LILCO to LIPA, a milestone—or rather, tombstone—

in the long saga of the plant. The plant was now owned by many of the same persons (on LIPA's nine-member board) who had been among its staunchest opponents: Richard Kessel, director of the New York CPB and longtime Shoreham foe; Irving Like, original opponent and counsel to the LHSG; Nora Bredes, executive coordinator of SOC and leader of demonstrations at the plant; and Thomas A. Twomey Jr., Easthampton attorney and counsel to Shoreham opponents.

On June 11, 1992, Shoreham set its final record, becoming the first commercial U.S. nuclear power plant ever to be dismantled. Once the NRC staff concluded that there were no significant environmental impacts associated with decommissioning, the commission determined no impact statement was necessary. In what was called a "symbolic reversal" of the traditional ribbon cutting, Governor Cuomo ignited a torch that cut a pipe to begin the demolition.[43]

With actual dismemberment of the plant underway, Shoreham–Wading River School District and SE2 at last gave up their struggle to keep Shoreham alive (or at least in suspended animation), ending nearly three years of unsuccessful legal challenges and court action. Their surrender followed a move by LIPA that harked back to an earlier incident (1985) in which LILCO had refused to pay property taxes on the Shoreham plant in order to pressure the county into being more cooperative. LIPA threatened to withhold $41 million in funds that were due Suffolk County, the town of Brookhaven, and the School District unless the parties (SE2 and School District) abandoned their litigation. As in the earlier instance, withholding of the funds could have caused the county's bonds to be reduced to "junk bonds," resulting in the equivalent of bankruptcy for the county. Looking noble in defeat, SE2 stated it would consider "its larger civic obligations" to county residents and withdraw all pending litigation.[44]

### Reactor Fuel, Only Slightly Used, Free to Good Home

Disposal of the eighty thousand cubic feet of low-level radioactive waste from the Shoreham plant proved another legal and logistical problem. LILCO had received immediate permission to ship the radioactive fuel supports to a low-level nuclear waste dump at Barnwell, South Carolina, amid concern that out-of-state shipments might be denied access to the Barnwell site by December 31, 1992. Still left at the plant was the radioactive fuel itself, which weighed about 100 tons and cost about $2.5 million a month to guard and maintain. There were 560 bundles of fuel assemblies in all. Each fuel assembly was 12 feet tall and 8 inches square and weighed about 2,000 pounds. "Each contained uranium pellets that are about a half-inch in length and quarter-inch in diameter [about the size of pencil erasers]. . . . There are 200 pellets per rod and 62 rods per assembly." Since their removal from the plant, the assemblies had been stored in a pool of water adjacent to the reactor. Because the fuel had been used for the equivalent of only two days at full power, most of the energy value of the fuel was still intact.[45]

In lieu of other alternatives, LIPA decided to send the used fuel assemblies to a French nuclear company (Cogema, Inc.) for reprocessing at a cost of about $74 million. This plan was vetoed by the Department of Defense as a threat to the national defense and security of the United States. LIPA then decided to pay the Philadelphia Electric Company $45 million to take the fuel for use in its two reactors near Pottsdown, Pennsylvania.

But how to move the fuel to Pennsylvania? The plan to ship it by tractor trailers to a rail siding and load the fuel into railcars was opposed by densely populated New York communities along the route. It was not until the end of 1993 that LIPA arranged to ship the fuel in oceangoing barges to Philadelphia, and thence by train to the Limerick power plant. Even at sea, Shoreham's fuel created controversy. The New Jersey Department of Environmental Protection and Energy sued in federal court to enjoin the barge shipment, arguing that the NRC had not analyzed the environmental impact on New Jersey's popular beaches should an accident occur that dropped reactor fuel anywhere near the land. The U.S. Court of Appeals for the Third Circuit avoided reaching whatever merit this claim might have had and affirmed dismissal of the case on procedural grounds.[46] The thirty-third and last shipment was made in June 1994, prompting Irving Like to crow: "We've exorcised the Devil." At the ceremony commemorating the final shipment of uranium fuel S. David Freeman, president and chief executive officer of NYPA, said soberly: "It was a damn fool idea ever to put a nuclear power plant on Long Island. . . . It's not really a happy occasion. There was that certain arrogance in the nuclear industry. We've forgotten now how completely oblivious to public safety the nuclear industry was."[47]

LIPA held a press conference to inform the public that all radioactive material had been removed. The NRC tested the site to confirm that it was free of dangerous levels of radiation and issued an order terminating Shoreham's license and releasing the site for "unrestricted use." The plant's opponents found it hard to believe that this day had finally arrived. Frank Jones, former deputy Suffolk County executive and formidable opponent of Shoreham, said: "When we first talked to our attorney he said the possibility of stopping the licensing of a nuclear power plant is less likely than stopping the rotation of the earth."[48] And, of course, Jones and company did not stop the licensing of Shoreham—just its operation.

Like every other cost connected with Shoreham, decommissioning costs escalated massively. Originally estimated to cost about $450 million (without including final disposal of the fuel), the cost rose to a range of $580–720 million in January 1991, including final property tax payments and disposal of the radioactive fuel. By the end of 1991, the bill for closing and decommissioning the plant—due to unanticipated property tax payments and delays in securing federal approval of the shutdown—was reported to exceed $1 billion. It was pointed out that LIPA would spend as much decommissioning Shoreham (which had operated for the equivalent of two days) as many utilities planned to spend on decommissioning reactors that had operated for forty years.[49]

It is unclear whether an alternate use can be found for the Shoreham property. Proposals have been made to convert the Shoreham plant into a variety of facilities, ranging from a reactor to produce tritium, a vital but short-lived constituent of nuclear warheads (the *New York Times*); a natural gas-fired plant (Representative Hochbrueckner, among others); the world's largest incinerator (Smith Barney, an investment firm); and a high-speed ferry terminal linking Long Island and Connecticut (Governor Cuomo).[50] The irony of making Shoreham itself the instrument for rapid evacuation from Long Island may have been lost on the proposer.

Conversion of Shoreham to a gas-powered plant, the proposal that received the most serious attention, was sidetracked when the CMS Energy Corporation of Jackson, Michigan (parent company of Consumers Power, the utility that serves most of Michigan), ran into financial difficulties. In 1991, LILCO announced that it had made an agreement with NYPA to build a $150 million electricity plant fueled by natural gas on LILCO-owned land in Holtsville, Long Island. The announcement appeared to rule out the possible conversion of Shoreham to natural gas, and LILCO said it would not need any new generating capacity for Long Island at least until the year 2005.[51] Thus the plant remains available.

## In Hindsight, They Didn't Need the Power

For years LILCO had claimed that Shoreham was necessary to sustain business expansion and new industry on Long Island and to meet electric demand during peak usage in the summer.[52] Without Shoreham, LILCO warned, Long Island would face brownouts, blackouts, the loss of new business, jobs, and economic growth; it would turn into an "economic wasteland." The New York State Energy Office and the CPB, on the other hand, claimed that LILCO could meet peak demand without Shoreham by better management of electric distribution, use of alternative sources of power, conservation programs, and increased purchases of electricity generated elsewhere. LILCO counterclaimed that the state studies overestimated conservation and underestimated the reserve power needed to assure reliability. A spokesman for Governor Cuomo said that LILCO had a record of "overstating" its electrical requirements for the summer and "discounting the saving potentials of conservation."[53] And so the arguments went on.

In 1988, with the abandonment of Shoreham imminent, LILCO changed its message. It asserted it could meet Long Island's energy needs without Shoreham by using aggressive conservation programs, purchase of power from other utilities, and installation of gas-fired turbines. One year later, regional energy experts described the New York region's power supply as "sufficient." Despite expectations of brownouts to reduce voltage or blackouts of service areas during peak summer periods, LILCO was able to meet service demand. It took old oil-fired units and kept them on-line, built a new oil-burning turbine plant on the Shoreham site, purchased power from neighboring utilities, and moved strongly into conservation and demand-side management (that is, reducing demand by making more efficient use of electricity).[54]

From New York's perspective, Shoreham's absence has not produced a deficit of electric power on Long Island. The PSC chairman Peter

Bradford noted, "LILCO agreed to stress load management, particularly peak load control in the short run . . . and has gotten 5 percent of their peak under load management." Bradford and Murray cannot resist pointing out the irony in all of this. It seems that DOE Deputy Secretary Moore characterized LILCO's load management alternative (gas turbines) as "baling wire and bubble gum" and yet unwittingly depended on LILCO when it sold power to the Washington, D.C., power pool during the hot summer months of 1989, when the Pennsylvania–New Jersey–Maryland (PJM) pool's nuclear units were unavailable. "Hence," says Murray, "while they were sitting there in Washington, making predictions [of blackouts], they were probably resting in the cool air conditioned comfort of electricity generated by LILCO."[55]

For the foreseeable future, due to conservation efforts, the high cost of electricity, the additional capacity from gas turbines, and the depressed state of the Long Island economy, there seems to be no problem in meeting Long Island's energy needs, albeit with more air pollution than if baseload electric generation came from nuclear power rather than fossil fuels. Governor Cuomo's energy aide noted that "Clearly one of the lessons is that we as a state, and as a society generally, can no longer afford to rely upon large-scale capital-intensive investments to meet our energy needs. We need a more diversified system, a more flexible system, one that relies upon smaller-scale less-capital-intensive investment."[56]

Peter Bradford questions whether demand-side management (which helped fill the gap on Long Island) would have been feasible had Shoreham gone into operation. He points out: "you can't implement an energy conservation program if half the customers hate the utility. Energy conservation requires a whole different kind of customer-utility relationship. The old-fashioned 'we generate the power and you consume it' relationship can't exist." There is, of course, some question, too, whether an operating Shoreham would have provided the expected power. LILCO's former president, an engineer, thought: "Shoreham was an overly complicated plant. With redundancy upon redundancy, it was uncertain that the plant could ever operate for a week without breaking down."[57]

# 8 The Message

Numerous policy issues are raised by NRC's behavior and actions during the Shoreham debacle. In large part, these deal with public trust and confidence in the nuclear regulatory body, essential ingredients of any hoped-for revival of the commercial nuclear power industry. Broader questions of trust in government are implicated.

While New York State and LILCO in Shoreham's aftermath have been battling over the terms of a possible state takeover, the federal government has taken steps it hopes will revive electrical utilities' long moribund enthusiasm for nuclear power. NRC has adopted new rules, supported by Congress and upheld by the courts, that are intended to streamline the licensing process and make it more "efficient and effective." The Energy Policy Act, enacted in October 1992, was a landmark victory for those in the nuclear industry who believe that the way to deal with public skepticism and fear is to throw the public out of the licensing process. As one of the act's supporters put it, the new law has a number of objectives, including the removal of "the regulatory barriers which have resulted in such colossal boondoggles as Shoreham."[1] The new procedures assure the nuclear industry that, once reactor construction is authorized, the public's chance to contest eventual operation of the plant will be limited.

Under the legislation, the NRC has adopted new rules that combine the reactor construction permit and operating license in a single "combined license" and has provided for only one hearing on safety issues, just prior to a plant's *construction*. Prior to operation, any further hearing is limited to the question whether the plant has been built in accordance with the acceptance criteria that were established prior to construction.[2] The public has no right to a further hearing on the adequacy of those criteria, even if new information arises that may cast doubt on crucial safety findings made at the early stage. Should safety issues arise during construction, a "10 CFR 2.206" petition (which raises a safety question at a particular nuclear plant) may be submitted to the NRC; this can be approved or denied according to staff and commission discretion, without a public hearing.

Proponents of the new rules believe that elimination of some of the NRC public hearings and curtailing the public's role in NRC proceedings will make licensing faster and more predictable. This is supposed to encourage the comeback of nuclear power. It is questionable, however, whether the NRC has learned an obvious lesson that the Shoreham episode should have taught it: contrived licensing of a nuclear plant the public does not trust can have results that are neither efficient nor effective.

As we have seen, Shoreham failed to operate because of intense public opposition, in which the governor played a key role, inspired in part by the utility's management incompetence and distrust of the NRC. Inefficiencies in the NRC licensing process were largely irrelevant to the outcome. The public by and large ignored NRC's findings and took the nonsafety of the plant for granted. Excluding safety issues from the OL proceeding would simply have reinforced that fatal distrust. Moreover, even if the new licensing reforms had been in place, the delays could have been equally long. Opponents of the plant would still have been entitled to some kind of hearing on whether the plant could comply with the new emergency planning rules (which were adopted long after Shoreham had received its construction permit), and the outcome would have been subject to judicial review. The same potential would have existed for protracted public litigation over the emergency planning controversy in one forum or another.

Can nuclear power be imposed on an unwilling public without its consent? If one believes that how we govern ourselves is more important then how we generate our electricity, one hopes the answer would be no. Usually, in our form of democracy, Americans value process more than results; if a law or policy or action has been arrived at by duly established processes in which people have confidence, most people most of the time will resign themselves to the result even though they may believe it to be wrong. In the Shoreham case, however, there seemed to be no confidence that the usual processes were either fair or impartial.

It is probably accurate to say that each side in the fight over Shoreham perceived the legal processes as leading inevitably to a result that was terribly wrong. The state, county, and general public believed that the NRC would find the emergency plan adequate, no matter what, and go on to license a plant they thought was dangerous. To block this, they engaged in noncooperation and what the commission called "bad faith obstructionism" in the licensing proceeding. For this the commission threw them out, an action that, of course, cleared the way for the license.

The NRC, on the other hand, could be accused of "bad faith facilitation." Once it saw that the state and county would not dutifully cooperate in emergency preparedness, and that it would be hard if not impossible to make the "reasonable assurance" finding required under a rational interpretation of NRC regulations, the commission began to stretch its rules out of shape in order to avoid the "wrong" result—that is, denial of the license for a plant the commission regarded as safe and desirable (as it regards almost all nuclear plants prepared for licensing). One might argue, then, that both sides showed a willingness to sacrifice fairness and rationality in the decision-making process as prescribed by law and regulations in order to prevent a substantive result that they felt would be a disaster.

The nuclear power controversy puts a greater strain than usual on the consensus that favors process over results because outwardly at least it is a fight over technical issues. Decisions made in this area typically reflect the assumption that experts are better able than the general public to deal with questions involving uncertainty and to determine whether the collective benefits of energy decisions outweigh the unknown risks.[3] But the public is no longer sure that technical expertise and specialized competence are being applied objectively in issues such as nuclear power. The notion that professionals and scientists alone possess the knowledge to represent the "public interest" in complex scientific controversies has been undermined by the suspicion that the expert may be promoting a not-so-hidden agenda, as well as by the increasing appearance of disagreements among the experts themselves.[4] Certainly, no politician with any degree of circumspection would take issue with the presumption that the public has a right to take part in all decisions affecting their interests, even those that require technical expertise. This may not make for the neat and orderly decision-making process that some professionals have in mind, but a process that remains open, accessible, and consistent with democratic principles is seldom entirely neat and orderly.[5]

From this perspective, if large numbers of people affected by a proposed nuclear power plant suspect that the utility's construction is faulty and doubt NRC's assurances of safety, as they did at Shoreham, then the outlook for operating that plant is questionable—or should be. Even LILCO's former president stated that he had no problem conceptually with the people deciding whether they wanted a nuclear plant, yea or nay. He just did not think "they should be allowed to make that decision after the plant was finished."[6]

The former NRC deputy general counsel Martin Malsch agrees, stating that he personally did not fault what Governor Cuomo did: "I might have done the same thing. If the people don't want a plant and they're willing to take the cost, it's their decision. They're the ones who will live with the plant and pay the electric rates." Malsch pointed out, however, that the commission's view of its licensing responsibilities does not take much account of public wishes. "The NRC's job is to license that plant, regardless of the opposition . . . to see whether it complies with the act and regulations. We'll listen to what other people say but their views aren't controlling. Ultimately it's our decision." (Legally, this is quite correct, but a skeptic might respond that at Shoreham the NRC seemed determined to make its regulations comply with the plant.) A cynical view from the other side is stated by LILCO's attorney Taylor Reveley, who believes that "most of the people will adjust very quickly to whatever evolves so long as . . . it serves their interest. The people are no more relevant to this sort of decision than any other and we don't in this country run by plebiscite on any issue."[7]

Former chairman Ivan Selin has described the NRC as "an agency that just has talked to itself," as one that did not engage in serious dialogue with the public.[8] He suggested that the NRC keep the public informed about what it does and why, "its strengths, its weaknesses, and the limitations of its role, vis-à-vis that of the licensees in assuring public health and safety." This has not, however, been a popular agency activity; perhaps out of fear of raising questions about a "promotional" role, the NRC hesitates to engage the public in a discussion of controversial issues. This is, of course, a no-win situation. Without direct communication and contact with high-level agency leaders, the public's concerns about emergency planning, environmental contamination, costs, risks, and safety problems—at Shoreham and elsewhere—go unanswered.[9]

Recognition of the value of two-way communication channels is of paramount importance in dealing with the outside world. When James Watkins, the former secretary of energy, advocated an aggressive approach to the industry "in communicating to what is unfortunately becoming more and more of a scientifically illiterate public the truth about nuclear energy, its efficiency, its environmental advantages, its future potential," he appeared to regard communication as a one-way process, in which the federal government "educates" the public about the benefits of nuclear power. The public is rightfully suspicious of such an approach.[10]

Instead of patronizing the public, the federal regulators would probably find increased candor and straightforwardness to be more effective for gaining public confidence. As former NRC Chairman John Ahearne advises, "Risk communication is a two-way street" and "openness is the surest policy." The Office of Technology Assessment, among others, has pointed out that "a concerted effort to identify and respond to the substance of critic's concerns could reduce acrimonious debate which contributes to negative public opinion." However, this does not always accord with the views of industry. Harold Finger, former president and chief executive officer of the U.S. Council for Energy Awareness, doubted that such an approach would be useful. "These skeptics don't control public attitudes," he wrote, "even though some of them have worked to generate unjustified public fears of nuclear energy. And there is little evidence, if any that they 'would be prepared to rethink their position.'"[11]

During the Shoreham-related years, NRC and LILCO tended to view the state and local politicians who disagreed with them as political opportunists who were more interested in their own jobs and continuation in office than accountable to their electorates in a positive sense. Perceptions still abound that the states are actors after the fact—as obstructionist, deterrent, or negative forces aiming to thwart national programs. LILCO's attorney believes, for example, that if the federal government were to say: "Nuclear power is in interstate commerce and we preemptively control the health and safety aspects, then the only thing the states could do is to block the facility at a certain location or find there's no need for it. Make those decisions up front and thereafter, it's totally a federal issue." John Sawhill, former deputy secretary of energy, shares the view that the federal interest is paramount and "the Shoreham debacle has taught Washington that it must override local objections to get new plants on-line." The *New York Times* editorialized that "only Washington has the technical expertise and the duty to weigh the national interest." And Howard Shapar, former executive legal director of the NRC, states that entry by the states into the nuclear power field is "unlawful, ill-advised, counter-productive to safety, and contrary to the national interest."[12]

The important point, it seems, is how broadly or narrowly one defines the "national interest." The health and safety aspects of a nuclear plant are greatest in its immediate area but they may range much further, as demonstrated by both Chernobyl and the controversy over where to dispose of high-level nuclear waste. The costs and benefits of the plant may be local, regional, or even international (reducing acid rain in Canada, for example). How are these diverse interests to be weighed against each other? Whatever the answer, it will involve more than simply declaring an overriding "national interest" in having nuclear power that sweeps away consideration for the views of those most directly affected by it. It seems likely that the states and localities would be more accepting of the potential burdens of having nuclear plants (or waste disposal facilities) in their backyards if they shared more rather than less responsibility in the licensing process, so that they were truly partners in the project rather than unwilling hosts. NRC's Harold Denton confirmed this sentiment, saying: "States are increasingly taking on important societal roles, and if we refuse to share any authority in this area it's going to be very tough to get people out there to accept it."[13]

The federal government is not likely anytime soon to share authority with the states over the safety of nuclear power plant design and operation. In the Atomic Energy Act of 1954, Congress reserved all authority over nuclear safety to the AEC, in part because at that time only that commission had the necessary scientific expertise. Many states now have staffs trained in radiological safety, and Congress has provided for state regulation of some nuclear materials and low-level radioactive waste. But few if any states could presently summon expertise in the more complex and controversial field of reactor safety that would come close to matching that of the NRC.

In the area of emergency planning, however, the states and the local governments can make a claim to relevant expertise to match or exceed that of the federal government. It may still be that the NRC knows best what the risks are of a radiological emergency at a nuclear plant that will require evacuation or sheltering, but the state and local governments would seem to know best their own abilities actually to carry out the needed emergency response.

Would a state be able to stop a plant from getting an operating license in the future because of emergency planning concerns? Taylor Reveley says definitively: "Shoreham solved the greatest threat to nuclear power in this country that had yet arisen. . . . [State and local vetoes of emergency planning] will never be raised again. They're dead." Bill Parler, former NRC general counsel, agrees it is "unlikely that a Shoreham-type situation could happen again" but cautions "after working in this area for more than three decades, I wouldn't rule out absolutely anything."[14]

Despite the NRC's efforts to reform and revise licensing procedures for the nuclear industry, current prospects for future nuclear power plants are not encouraging. New designs for smaller and simpler reactor systems are on the drawing boards, but critics say further research is needed. Several of these new reactor designs use "passive" safety systems, which would rely on natural forces rather than active mechanical devices to keep a reactor from going out of control. Nevertheless, according to a 1992 report of the National Research Council (research arm of the National Academy of Sciences), efforts to build more nuclear plants will fail unless answers can be found to questions of reduced demand, high cost of construction and operation, regulatory and financial uncertainty, waste disposal, and safety concerns. (Notwithstanding the economics, some observers are optimistic about the future of nuclear power. Alvin Weinberg, for example, believes with the industry that there will be a second nuclear era in America, "one that embodies improvements based on our experience in the first nuclear era.")[15]

Even if these questions are answered to the satisfaction of technical experts, there will remain a need for the NRC to act—and to be perceived by the public—as a bold, vigorous regulatory body, with an aggressive approach to regulatory standards, safety analyses, quality assurance, and investigations of allegations of whistle-blowers (that is, those with safety concerns within and outside the agency), with ends of public health and safety paramount to all other considerations. In short, re-regulation, instead of de-regulation.

NRC chairmen have not necessarily taken a sophisticated approach to building public confidence in the agency's approach to safety concerns. Chairman Lando W. Zech Jr. stated, for example, that public confidence in safe nuclear power would emerge as a by-product of "the regulatory authority and the industry operating with discipline, and together working in a responsible and effective manner." His successor, Admiral Kenneth Carr, asserted that the public could have confidence in the safety of reactors because "their operators were patriotic Americans." Closer to the mark, perhaps, former NRC chairman Ivan Selin said the only way to restore public confidence in the nuclear technology is "to insist on safety, safety and more safety."[16] A plan to make the NRC a more responsive and "user-friendly" agency includes such activities as public workshops and providing the public with speedier and more complete access to information about NRC decisions, as in press briefings, telephone communications, freedom of information requests, and electronic information exchanges. But recent events at Northeast Utilities' Millstone plant highlight the NRC's continued failure to enforce its own rules and give a skeptical public continued reason to doubt whether the agency insists on safety first.[17]

Shoreham should serve as a warning against limiting the public's ability to raise safety-related issues. All the effort of building and licensing Shoreham went for naught because the process reinforced—rather than abated—distrust that the agency would make a fair or correct finding. The NRC's assurances about the safety of Shoreham were consciously and openly rejected, even by the highest state government officials. Unless this distrust is cleared up, NRC's effort at licensing reform may cause more problems for nuclear power than it solves. If a concerned and skeptical public refused to believe NRC safety findings under current law, reducing the opportunities for public hearings will only increase the distrust. Streamlining the licensing process will be counterproductive if it detracts further from public confidence in the commission's actions.

To LILCO's lawyer, the Shoreham story established that, "confronted with the most powerful, hideous, relentless constellation of opposition, you can get licensed."[18] Yes, the NRC did confirm that it would make licensing decisions in defiance of public opinion, out of either courage or contempt. But what good did that do at Shoreham? Shoreham shows that the mere process of issuing a license does not assure that a nuclear power plant will come on-line. Shoreham also shows how an agency that appears to be trying too hard to overcome obstacles to licensing undermines the credibility of the process. Scrapping Shoreham after the license issued was in effect a public vote of no confidence in the NRC's findings. NRC should have realized that its actions invited skepticism about its neutrality as a nuclear regulator. Moreover, the NRC's accommodating approach to LILCO's emergency planning problems may have encouraged the utility to keep on squandering resources that might have been saved if the whole project had been scrapped earlier.

The prospects for reviving nuclear power will surely not be advanced if another plant is built, licensed, and then promptly junked primarily because people are afraid of it. This fear may be groundless but faster licensing with reduced public participation is hardly likely to convince the public of this. NRC's new "combined license," with its emphasis on limiting public input as much as possible to the construction permit stage (that is, focusing public participation where it can do the least damage) sends an unmistakable message that the agency and the nuclear industry have not begun to trust the people, not when the people might cause delays in reactor licensing. The distrust will almost certainly be reciprocated. The next "reform" needs to be a more serious commitment to building mutual confidence between the nuclear regulators and the public that has to live with the plants they license. This should have come first.

# Epilogue

## There Is Just No End to the Shoreham Debacle

One might think that getting rid of the plant would bring an end to the long-drawn-out Shoreham controversy, but this was not so.[1] Shoreham's continuing legacy has been dissatisfaction with LILCO's electric rate structure, the highest in the nation. In 1990, LILCO's customers paid an average of 11 cents per kilowatt hour of electricity, topping Consolidated Edison Company of New York, the previous holder of this dubious record. In 1991, LILCO received an actual 23.2 percent increase in electric rates for the first two years of the settlement—instead of the 10 percent customers might have expected to pay—because of higher fuel costs and some other expenses. LILCO warned that it would continue to raise future rates more than the 5 percent per year allowed by the settlement if its expenses went up or its sales went down.[2]

Then, in a sharp departure from past expectations, because of lower interest rates and oil costs and additional purchases of cheaper power from outside Long Island, LILCO announced it would seek a rate freeze in its base rates in 1995 and 1996 followed by only a 4 percent increase in 1997. Customers' bills rose just the same because of fuel adjustments and lower-than-expected electric sales. At last count, LILCO charged an average of 16.8 cents a kilowatt hour, nearly twice the national average of 8.8 cents and well above the rates of Con Edison and utilities in the New York metropolitan area.[3]

Because of the high rates, a public takeover of Shoreham is still on many people's minds. Initially, a couple of LIPA trustees called for a takeover study, a proposal adopted by Governor Cuomo during a fierce gubernatorial election battle in 1994. Cuomo maintained that a state takeover would lower electric rates for LILCO customers by about 10 percent, because of the elimination of shareholder dividends and federal taxes and lower capital costs. Although Wall Street's reaction was to call the proposal "more political bravado than substance," many people, including some LILCO executives, took it seriously. Because of recent congressional legislation to foster competition in electric markets, power companies were beginning to bid against other suppliers in meeting the demand for power, and LILCO feared it might lose its monopoly stronghold on Long Island. The *New York Times* was quick to call the takeover proposal "an election-year ploy" but nevertheless thought that the potential savings might be great. Gubernatorial candidate Pataki disputed the potential savings but pledged to lower LILCO's rates.[4]

Following his election, Governor Pataki came out against a takeover proposal, dismissing it as politically irrelevant, "irresponsible," and showing a "blatant disregard for taxpayers." LILCO turned it down as well. Three weeks later, the Suffolk County executive—a major Republic official and friend of the governor—endorsed the proposal, saying it was the best one he had seen for reducing the high electric rates. In an eighteen-to-one vote, the county legislature endorsed it too. In a 180-degree turn from his previous position, Governor Pataki embraced the dissolution of LILCO as a way to reduce electricity rates on Long Island. It is not irrelevant to point out that Governor Pataki won his largest margin of victory in the 1994 election in Suffolk County—fifty-four thousand votes, which constituted about a third of his total margin of victory. As with Governor Cuomo, Suffolk is an important jurisdiction in a New York governor's political universe.[5]

The public takeover proposal could erase Suffolk County's liability for a pending lawsuit with LILCO. At the time of the New York–LILCO settlement agreement, a suit was pending by LILCO against Suffolk County, the town of Brookhaven, and the Shoreham–Wading River School District, charging that the Shoreham plant was overassessed between 1976 and 1984. LILCO argued that the plant had no economic value at any time during its history and therefore should not be taxed. State supreme court judge Thomas M. Stark ruled in October 1992 that the plant had been overassessed and ordered Suffolk County to pay $55 million in tax refunds for the years 1976–1983. Because of interest payments, the penalty rose to $83 million by 1995. This has been largely repaid.

In a recent court case, LILCO sought additional tax refunds of up to $500 million, plus interest for the years 1984–1992 and payments made in lieu of taxes for 1992–1995. Judge Stark ruled on November 8, 1996, that once the New York–LILCO settlement agreement was signed in 1989, the plant no longer had taxable value. Yet the county and local governments continued to overtax the plant (even while they were doing everything they could to shut it down) "as if its operation were a foregone conclusion." His decision called for the county, the town, and the school district to pay LILCO as much as $1.16 billion in refunds and interest for overtaxing the Shoreham plant, amounts that would "bankrupt the school district and saddle the local governments with crushing tax increases."[6]

These potentially devastating financial consequences have increased pressure on Governor Pataki for a state takeover of LILCO that would include a negotiated settlement of the tax suit. If Judge Stark's judgment is allowed to stand it would be a mixed benefit for ratepayers; although LILCO has vowed to use the proceeds to reduce electric rates, the judgment could result in enormous property tax increases for the governments liable for paying the refunds. Following the ruling, the Suffolk County executive, Robert Gaffney, said: "We've had a glimpse of doomsday." Rick Brand of *Newsday* suggested that Gaffney might have been referring to his own political career—and that of Governor Pataki, as well as to Suffolk taxpayers. (Irving Like observed: "You could use the Shoreham plant as a retirement residence for all the people who were either elected or defeated because of it.")[7]

It is not "doom" for everyone, though; LILCO and President Catacosinos appear to have more lives than a cat. At the end of 1996, LILCO announced that it would merge with Brooklyn Union Gas Company, a company serving over 1 million customers in the outer boroughs of New York City. The transaction, which is subject to approval by shareholders and regulators, would create a new company with about

2.2 million customers. It is expected to benefit customers of both companies by saving $1 billion over the next ten years by reducing the combined work force and streamlining operations. The arrangement calls for Dr. Catacosinos to remain as chairman and chief executive, at double his present salary of $600,000 per year. (He would also receive payment of $6 million when the merger is completed.) The ever-vigilant Like points out that LILCO is coming out ahead once again: the merger "handsomely enriches LILCO stockholders, bondholders and management, while depriving ratepayers of desperately needed rate cuts."[8]

The merger would remove LILCO as an independent company, thereby fulfilling one of Governor Pataki's promises to the residents of Long Island. Pataki still has to make good on reducing the electric rates, though. The merger would not directly deal with the problems of LILCO's debt burden of nearly $5 billion for Shoreham or the $1.1 billion debt owed to LILCO for excess taxes paid for the plant. Presumably, these problems would be addressed by a purchase by LIPA of part or all of LILCO's electric system and the use of tax-exempt bonds to refinance LILCO's Shoreham debt. (When this happens, LILCO shareholders will receive a bonus worth more than $2 a share.) In March 1997, Governor Pataki announced the details of a LIPA purchase subject to the approval of the state Democratic and Republican leaders, the PSC, and the Internal Revenue Service, among others. In line with past Shoreham excesses, the agreement involves the sale of more than $7 billion in municipal bonds, the largest municipal bond issue in history.[9]

Pataki's plan will not magically reduce the cost to society of building and then abandoning a $6 billion nuclear plant. All that will happen is that the public's cost of serving the debt will be reduced by refinancing with tax-free bonds. This will not create new money to go into LILCO's coffers or anyone else's. Rather, it will just take money from all federal taxpayers and give it to Long Island ratepayers. The deal has received less than universal acclaim from the Long Island delegation in the state legislature, the Suffolk County legislature, and some consumer groups who believe it is overly generous to LILCO stockholders and Wall Street investment advisers. If Governor Pataki's actions result in a takeover of LILCO, LILCO's stockholders should get a fair value for their shares—but no more—and LILCO's management should not receive huge bonuses. Considering how well they have worked around the Shoreham disaster, maybe the management would deserve such rewards, but one hopes they will not be paid out of the pockets of the taxpayers and the ratepayers.

# Appendix
## Interviews

J oan Aron conducted all of the interviews. Leo Slaggie accompanied her on about twelve interviews, or roughly one-third of the total. All interviews were audiotaped except those marked "by telephone" and the interview with Charles Pierce, former LILCO board chairman, who did not wish to be taped. Joan Aron transcribed all the tapes, which are now in her possession. Interviewees were selected on the basis of their knowledge of and participation in the Shoreham debacle and also on the recommendations of those already interviewed.

Per Alin. Former professor of ancient history, State University of New York, Stony Brook. November 7, 1991.

Frederick M. Bernthal. Former NRC commissioner, 1983–1988. April 19, 1993.

John C. Bierwirth. Former chairman and chief executive officer, Grumman Aircraft. June 10, 1992.

Joel Blau. Staff member, New York Consumer Protection Board. March 31, 1992.

Bernard M. Bordenick. NRC attorney. March 3, 1993.

Peter A. Bradford. Former chairman, New York Public Service Commission, and NRC commissioner, 1977–1982. January 9, 1992.

Herbert Brown. Former attorney for Suffolk County. By telephone, December 2, 1991.

Karen Burstein. Former member, Marburger Commission and New York Public Service Commission; former president, New York State Consumer Protection Board. April 1, 1992.

Matthew C. Cordaro. Former senior vice president of operations and engineering, LILCO. June 11, 1992.

Harold Denton. Former director, Office of International Programs, Office of Governmental and Public Affairs, and Nuclear Reactor Regulation, NRC. April 12, 1993.

Thomas Downey. Former member, U.S. House of Representatives. February 18, 1992.

Spiros Droggitos. Former assistant to NRC commissioner James K. Asselstine. March 3, 1993.

Darrell Eisenhut. Former NRC project manager, Shoreham. March 13, 1993.

Anthony Fainberg. Former staff member, Office of Technology Assessment. By telephone, April 4, 1991.

Ira L. Freilicher. Attorney, Hunton and Williams, New York, N.Y.; former vice president of Laws and Corporate Affairs, LILCO. April 1, 1992.

George Hochbrueckner. Former member, U.S. House of Representatives. December 11, 1991.

Frank Jones. Supervisor, Town of Islip; former deputy county executive, Suffolk County. December 12, 1991.

Richard Kessel. Trustee and former chair, Long Island Power Authority; former executive director, New York State Consumer Protection Board. March 1, 1992.

Charles Komanoff. Director, Komanoff Energy Associates. November 7, 1991.

Lee E. Koppelman. Executive director, Long Island Regional Planning Board; director, Center for Regional Policy Studies, State University of New York, Stony Brook. November 7, 1991.

Herbert J. Kouts. Chairman, Department of Nuclear Energy, Brookhaven National Laboratory, N.Y. December 3, 1991.

Irving Like. LHSG attorney and special counsel to Suffolk County; LIPA member. December 12, 1991.

Martin G. Malsch. Former deputy general counsel, NRC. February 9, 1993.

Joseph W. McDonnell. Senior vice president for marketing and external affairs, LILCO. September 20, 1993.

Frank Murray. Deputy secretary to Governor Cuomo; former commissioner, New York State Energy Office. January 9, 1992.

Fabian Palomino. Special counsel of Governor Cuomo and New York State in Shoreham proceedings. December 13, 1991.

William Parler. Former general counsel, NRC. March 22, 1993.

Charles R. Pierce. Former chairman and chief executive officer, LILCO. December 13, 1991.

C. William Reamer. NRC attorney. March 22, 1993.

W. Taylor Reveley III. Attorney, Hunton and Williams, Richmond, Va.; principal attorney for LILCO during Jamesport hearing and Shoreham operating license proceeding. January 14, 1993.

Vance Sailor. Former nuclear physicist, Brookhaven National Laboratory, and head of Suffolk Scientists for Cleaner Power and Safe Environment. December 11, 1991.

Richard Schuler. Former member, New York State Public Service Commission. By telephone, November 12, 1991.

Miro M. Todorovich. Executive director, Scientists and Engineers for Secure Energy, Inc. November 8, 1991.

Matthew L. Wald. Reporter, *New York Times*. By telephone, May 15, 1991.

Jerry N. Wilson. Former NRC project manager, Shoreham. By telephone, March 11, 1993.

Wilfred Uhl. LILCO president, 1978–1984. November 16, 1992.

# Notes

ALAB    Atomic Licensing Appeal Board

ASLBP   Atomic Safety and Licensing Board Panel

CFR     Code of Federal Regulations (e.g., 10 CFR 50.47 refers to vol. 10, pt. 50, sec. 47)

CLI      Commmission Licensing Issuance

EDNY    Eastern District New York (for U.S. District Court)

FR       Federal Register

HQ      Headquarters

LBP     Licensing Board Panel

NDNY   Northern District New York (for U.S. District Court)

NR&C    *News Reviews and Comments* (internal NRC publication)

PDR     Public Document Room

SEC'Y    Office of the Secretary (refers to official papers for Commission consideration)

TR       Transcript

## Chapter 1. A World-Class Fiasco

1. Cambridge Energy Research Associates, *U.S. Public Opinion and the Electric Power Industry: Challenges of Competition* (1995), p. 11, and *Special Report: Fueling the Race for the Presidency* (1992); Lou Harris, "Poll on Nuclear Energy," National Public Radio, "Morning Edition," December 29, 1988.

2. Letter from Jan Beyea, senior staff scientist, National Audubon Society,

*Issues in Science and Technology* (spring 1990), p. 9; Jan Beyea, *National Journal*, September 9, 1989, p. 2199. See also U.S. Nuclear Energy Industry, *Strategic Plan for Building New Nuclear Power Plants: Fifth Annual Update* (November 1995); Alvin M. Weinberg, *The First Nuclear Era: The Life and Times of a Technological Fixer* (New York: American Institute of Physics, 1994), p. 279; "Group Sees Brighter Future for Nuclear Power Industry," *Central Penn Business Journal*, April 19, 1996.

3. *Nuclear Power: Technical and Institutional Options for the Future* (Washington, D.C.: National Academy Press, 1992), p. 57.

4. The Vietnam War in the late 1960s triggered a distrust for a strong federal government, which was then nourished during the Reagan and Bush years and which intensified in the 1990s. See, for example, Susan J. Tolchin, *The Angry American: How Voter Rage Is Changing the Nation* (Boulder, Colo.: Westview Press, 1996). See also the final report of the Secretary of Energy Advisory Board Task Force on Radioactive Waste Management, "Earning Public Trust and Confidence: Requisites for Managing Radioactive Waste," June 1993, esp. p. 11. The report points out, *inter alia*, that "a high level of trust and confidence buttresses the legitimacy of action in the public sphere. Conversely, a low level erodes that legitimacy and calls into fundamental question the bond between those who govern and those who are governed" (p. 2).

5. Remarks attributed to Richard H. Bornemann, lobbyist for the American Nuclear Energy Council, in John Rather, "Utilities Study the Lessons of Shoreham," *New York Times*, March 16, 1986; Sarah Lyall, "Federal Official Sees Bid to Save Shoreham Plant," *New York Times*, June 16, 1989; Hank Boerner, "Shoreham: Long Island's Phoenix Rising from the Ashes," *New York Tribune*, October 24, 1990; Thomas M. Lippman, "Nearly New A-Plant Going to Scrap Heap," *Washington Post*, June 18, 1992. Secretary Watkins cited in Victor Kirk, "The Nuclear Turn On," *National Journal*, September 9, 1989, p. 2197; Peter Bradford, "Call Me Ishmael: Reflections on the Role of Obsession in Nuclear Energy Policy," Address to Annual Convention of the National Association of the Regulatory Utility Commissioners, Boston, November 13, 1989, p. 7; editorial, *New York Times*, April 12, 1989.

6. Matthew L. Wald, "At Shoreham Plant: Ready with Foreboding," *New York Times*, May 1, 1988. See also "The Shoreham Saga: Twenty Years of Uncertainty," *IEEE Spectrum* (November 1987), pp. 24–37; *New York Times*, July 16, 1985, and May 1, 1988. A series of articles by Stuart Diamond, "Shoreham:

What Went Wrong," *Newsday*, November 15–22, 1981, is useful in providing insights on the Shoreham experience prior to 1981. See also Peter Marks, "Eye of the Storm," *Newsday*, May 15, 1988, pp. 93–113.

7. Secretary of Energy James Watkins, *Inside Energy*, May 1, 1989, p. 3.

## Chapter 2. Enthusiasm and Errors

1. *New York Times*, March 16, 1994. See also Robert C. Wood, *1400 Governments: The Political Economy of the New York Metropolitan Region* (Cambridge, Mass.: Harvard University Press, 1961).

2. Interview with John C. Bierwirth, former chairman and chief executive officer, Grumman Corporation, June 10, 1992. In 1996, LILCO served a population of 2.7 million in Nassau and Suffolk Counties and the Rockaways in Queens.

3. Interview with W. Taylor Reveley III, LILCO lawyer during Shoreham and Jamesport proceedings, January 14, 1993.

4. The famous characterization of the nuclear option as a Faustian bargain originates with Alvin Weinberg, former director of the Oak Ridge National Laboratory, in "Social Institutions and Nuclear Energy," *Science* 177, July 7, 1972, pp. 27–34. Generally supportive of nuclear power, Weinberg nevertheless acknowledged the trade-off between the expected benefits and the potential risk of accidents and problems of waste disposal.

5. *New York Times*, January 6, 1969, February 17, 1971. See also J. Samuel Walker, *Containing the Atom: Nuclear Regulation in a Changing Environment, 1963–1971* (Berkeley and Los Angeles: University of California Press, 1992), pp. 9–12.

6. New York State Atomic Space and Development Authority, *Seventh Annual Report, April 1, 1968, through March 31, 1969: Ten Years of Atomic Progression in New York State*, p. 7; *Journal of Commerce*, April 3, 1956.

7. State law in this area was preempted by the Atomic Energy Act. See *Northern States Power Co. v. Minnesota*, 447 F. 2d 1143 (8th Cir. 1971), *aff'd*. 405 U.S. 1035 (1972); *Pacific Gas and Electric Co. v. State Energy Resources Conservation and Development Commission et al.*, 461 U.S. 190 (1983). Only the AEC could regulate the radiological safety of nuclear plants and facilities that fabricated or reprocessed nuclear fuel.

8. New York State Public Service Commission (PSC), *Annual Report, 1968*, p. 35.

9. *New York Times,* October 28, 1969.

10. Interview with Frank Jones, former deputy county executive, Suffolk County, December 12, 1991; *New York Times,* April 22, 1965, p. 1.

11. Ibid.

12. Interview with Wilfred Uhl, former president, LILCO, November 16, 1992.

13. Dennison's remarks come from *Newsday,* April 13, 1966, quoted in Karl Grossman, *Power Crazy* (New York: Grove Press, 1986), pp. 92–93; LILCO press release, April 13, 1966; Dominey, quoted in *New York Times,* April 14, 1966; interview with Lee A. Koppelman, executive director, Nassau-Suffolk Regional Planning Board, November 7, 1991.

14. Interview with Vance Sailor, physicist, Brookhaven National Laboratory, December 11, 1991.

15. Koppelman interview.

16. "Pre-hearing Conference," Docket 50-322, March 13, 1970, p. 13 (in HQ PDR); "The Shoreham Saga," *IEEE Spectrum* (November 1987).

17. *Newsday,* January 2, 1983; interview with Ira L. Freilicher, former vice president for Law and Corporate Affairs and vice president for Public Affairs, LILCO, April 1, 1992.

18. Sugden's memorandum was later used as an exhibit in Case No. 27563 developed by the New York PSC in 1985, which penalized LILCO for mismanagement of construction activities.

19. "The Shoreham Saga," *IEEE Spectrum* (November 1987).

20. Freilicher interview; interview with Matthew C. Cordaro, former senior vice president, Operations and Engineering, LILCO, June 11, 1992.

21. Reveley interview. See also Cordaro interview; *IEEE Spectrum* (November 1987).

22. See Section 189a of the Atomic Energy Act of 1954, as amended 42 USC 2239a.

23. New York Public Service Commission, "An Evaluation of Long Island Lighting Company's Major Project Management Process," study prepared by Booz, Allen, and Hamilton, August 25, 1978, pp. 1–4 (in HQ PDR). Steven Ebbin and Raphael Kaspar characterize the AEC hearing process in the early 1970s as "nothing more than a charade, the outcome of which is for all intents and purposes predetermined" (*Citizen Groups and the Nuclear Power Controversy: Use of Scientific and Technological Information* [Cambridge, Mass.: MIT Press, 1974], p. 246).

24. Reveley interview. In this respect they were little different from the hearings conducted by regulatory bodies in general. For example, a study conducted by the U.S. Senate Committee on Governmental Affairs in the 1970s found little interest in regulatory proceedings and little or no organized public interest participation, stating: "In more than half of the formal proceedings, there appears to be no . . . organized public interest participation whatsoever and virtually none at informal agency proceedings. In those proceedings where participation by public groups does take place, typically it is a small fraction of the participation by the regulated agency. One tenth is not uncommon; sometimes it is even less than that" (*Study on Federal Regulation: Public Participation in Regulatory Agency Proceedings* 3 [July 1977]: p. vii).

25. Reveley interview. See also *New York Times*, January 24, 1971.

26. *Newsday*, January 2, 1983; Reveley interview.

27. Interview with Martin G. Malsch, NRC deputy general counsel, February 9, 1993.

28. Interview with Irving Like, attorney for the Lloyd Harbor Study Group, December 12, 1991; Irving Like, "Multi-Media Confrontation—The Environmentalists' Strategy for a 'No-Win' Agency Proceeding," *Ecology Law Quarterly* 1 (1971), pp. 495, 508. See also *New York Times*, September 20, 1970, January 24, March 21, 1971.

29. Sailor interview; *Nuclear Industry* 18 (June 1971), p. 23.

30. Malsch interview.

31. Reveley interview.

32. Malsch interview.

33. *Newsday*, January 27, 1983; *New York Times*, October 7, 1971.

34. Malsch interview.

35. Ibid.

36. *Calvert Cliffs Coordinating Committee Inc. et al. v. United States Atomic Energy Commission*, 449 F. 2d 1109 (D.C. Cir. 1971).

37. Additionally, because the AEC hearings on Shoreham were ongoing when New York State's new water quality standards went into effect, the New York Department of Environmental Conservation held environmental hearings from August to December 1971, concerning the water quality and the aquatic impact of Shoreham on fish eggs and larvae in Long Island Sound.

38. Statement of Jack M. Campbell, ASLB chairman, transcript, Docket 50-322 T17, December 2, 1970, p. 2509, and T16, December 1, 1970, p. 2306 (in HQ PDR).

39. Sailor, quoted in *Nuclear Industry* 18 (June 1971), p. 25.

40. Malsch interview.

## Chapter 3. Constructing Shoreham and Making Enemies

1. The words of the heading are taken from the interview with Karen Burstein, former New York PSC member, April 1, 1992. LILCO officials are cited from *Newsday*, November 16, 1981.

2. Such a major redesign would probably have required an amendment to LILCO's CP application, opening the possibility of still more controversy and delay at the hearing. To make a major design change during construction, LILCO would have had to apply for an amendment to its permit, and the intervenors would have been entitled to a whole new hearing. One can see why LILCO might reasonably have decided to get by as best it could with the old design.

3. New York PSC, Case No. 27563, Opinion 85-23, "Opinion and Order Determining Prudent Costs," December 16, 1985, pp. 75–82.

4. Cordaro interview; Freilicher interview; Cordaro interview.

5. New York PSC, Case No. 27563, pp. 33–38.

6. The Energy Reorganization Act of 1974 abolished the old AEC and transferred its regulatory authority over nuclear power to the newly created Nuclear Regulatory Commission (NRC). The point of the act was to create a new regulatory commission that did not have the conflicting assignment of promoting the industry it was supposed to regulate. The AEC's promotional and developmental responsibilities were vested in an Energy Research and Development Administration, which subsequently became absorbed in a new Department of Energy.

7. Cordaro interview.

8. Uhl interview.

9. Malsch interview; James Cook, "Nuclear Follies," *Forbes* (February 1985), p. 82.

10. Harold Denton, quoted in *Newsday*, November 18, 1981, p. 7. See also New York PSC, Case No. 27563, pp. 70–75.

11. Telephone interviews with Jerry N. Wilson, March 11, 1993, and Darrell Eisenhut, March 13, 1993. Quote is from Wilson. See also New York PSC, Case No. 27563, 85–86.

12. Cordaro interview; New York PSC, Case No. 27563, 38–41.

13. Interview with Joel Blau, CPB staff member, March 31, 1992; interview with Frank Murray, energy advisor to Governor Cuomo, January 9, 1992.

14. Uhl interview.

15. Interview with Harold Denton, former director, Nuclear Reactor Regulation, NRC, April 12, 1993.

16. Interview with Bernard M. Bordenick, NRC attorney, March 3, 1993; Freilicher interview.

17. Stuart Diamond, "Shoreham: What Went Wrong?" *Newsday*, November 17, 1981. See also *Newsday*, March 8, 1979, and *New York Times*, May 23, 1982, quoting a 1978 state productivity study; Bob Wiemer, "A Mad, Mad, Mad, Mad Power Plant," *Newsday*, December 11, 1988.

18. Uhl interview; Cordaro interview.

19. Eisenhut interview; Cordaro interview.

20. Denton interview.

21. The words of the heading are taken from the interview with Thomas Downey, former member of the U.S. House of Representatives, February 18, 1992.

22. *Newsday*, November 16, 1981, p. 24.

23. Uhl interview. He added that the company never perceived in those days that public opinion would come to be the "dooming factor." See also New York PSC, Case No. 27563, appendix A, "History of the Shoreham Project"; *New York Times*, December 4, 1988.

24. Freilicher interview.

25. New York PSC, Case No. 27563, pp. 5, 7–8, 31, 32, 50, 58, 72, 88.

26. About $95 million of this total amount was attributable to LILCO's imprudence with respect to the diesel generators, a problem that developed in 1983 as the plant was nearing completion. Three PSC commissioners dissented from the majority decision regarding the portion of the adjustment related to the diesel generator failure, advocating a higher disallowance of $1.9 billion for LILCO's imprudence. See dissenting opinion to New York PSC, Case No. 27563.

27. Quote in heading is from Reveley interview.

28. State regulation of nuclear-power-plant radiological safety was preempted by the Atomic Energy Act, but the act left the states free "to regulate activities for purposes other than protection against radiation hazards." 42 USC 2021 (k).

29. Quotes from Frances Cerra, "A Genial Gadfly vs. LILCO," *New York Times*, August 7, 1977; *Newsday* also described the hearings on May 5, July 7, August 2, 1977.

30. Cerra, "A Genial Gadfly vs. LILCO"; Like interview.

31. Quotes from *Newsday*, May 16, 1978, and July 2, 1979. See also *Newsday*, August 24, 1977, December 28, 1978, and July 24, 1979.

32. *New York Times*, July 20, 1979, p. 1, and January 29, 1980. *Newsday* reported that the only persons disappointed with the decision were union officials, because of the four thousand or so construction jobs lost (January 29, 1980).

33. Freilicher interview.

34. Reveley interview.

35. Freilicher interview.

36. Cordaro interview.

37. The alternative to once-through cooling is a system that recycles the water used to cool the heat exchangers. This is done by the famous cooling towers that for many people are the hallmark of nuclear power. The cooling water passes over the heat exchangers, condensing the steam inside the exchangers, and is itself converted into steam. The steam rises inside the cooling tower, and as it rises it cools and condenses—in effect transferring the heat to the atmosphere—and falls back into the bottom of the cooling tower to be used again for cooling the heat exchangers. Hence "twice-through" or "many-times-through" cooling.

38. Freilicher interview; Burstein interview

39. Reveley interview.

40. Freilicher interview; Cordaro interview.

41. Interview with Charles Komanoff, independent consultant on economics and energy *(inter alia)*, November 7, 1991.

42. *New York Times*, March 28, 1976; interview with Richard Kessel, former New York CPB executive director and former LIPA chairman, April 31, 1992.

43. Uhl interview.

44. Quotes from *Newsday*, November 20, 1981; see also *Newsday*, May 15, 1988.

45. Frances Rivett, quoted in *Newsday*, April 16, 1980. By 1985, *Forbes* reported that the cost rose to a "grotesque $5,192 a kilowatt for LILCO's Shoreham plant" (February 11, 1985, p. 83).

46. *Newsday*, April 16, 1980, November 20, 1981, March 12, 1982.

47. Komanoff interview; *Newsday*, April 4, 1979.

48. Peter Cohalan, quoted in *Newsday*, April 21, 1982; Cordaro interview.

49. Kessel interview.

50. *Newsday,* April 2, 1979.

51. Interview with Fabian Palomino, special counsel of Governor Cuomo and New York State in Shoreham proceedings, December 13, 1991.

52. Bierwirth interview.

**Chapter 4. Three Mile Island**

1. The facility at TMI was a pressurized-water reactor, a different design from the GE boiling-water reactor being constructed at Shoreham. Both reactors, however, are so-called light-water reactors, by far the predominant U.S. design. All U.S. power reactors—except Fort St. Vrain in Colorado (no longer in operation)—rely on water to cool the reactor core and to moderate the neutron flux, that is, to slow down the neutron produced by nuclear fission to increase the probability that the neutrons will initiate another energy-releasing fission reaction. Fort St. Vrain was a high-temperature gas-cooled reactor.

2. See NRC Special Inquiry Group, *Three Mile Island: A Report to the Commissioners and to the Public* (Washington, D.C.: GPO, 1980), for a detailed discussion of the accident sequence at TMI Unit 2.

3. *In re. TMI Litigation Consolidated Proceeding,* U.S. District Court Middle District of PA 227 F Supp 834 (1996). Judge Rambo granted summary judgment for the defendant utility (GPU), throwing out a suit by citizens living in the area for damages they assertedly sustained as a result of the accident.

4. See 10 CFR, 50, App. E (1971).

5. U.S. Congress, House, Joint Committee on Atomic Energy, *Report on Bill to Amend Price-Anderson Act to Provide Immediate Assistance to Claimants and to Require Waiver of Defenses,* 89th Cong., 2d sess., H.R. Res. No. 2043, 1966; and U.S. Congress, Report by the House Committee on Government Operations, *Emergency Planning Around U.S. Nuclear Power Plants: Nuclear Regulatory Commission Oversight,* 96th Cong., 1st sess., H.R. Rep. No. 96-413, 1979, p. 4.

6. Malsch interview.

7. See David Comey, "Do Not Go Gentle into That Radiation Zone," *Bulletin of Atomic Scientists* (November 1975); Natural Resources Defense Council, *Letters and Comments on NRC Proposed Emergency Planning Amendments,* October 23, 1978. See also U.S. Comptroller General's Report to the Congress, *Areas Around Nuclear Facilities Should Be Better Prepared for Radiological Emergen-*

*cies*, EMD-78-110, March 30, 1979; U.S. Congress, 96th Cong., 1st sess., H.R. Rep. No. 96-413, August 8, 1979; Report of the President's Commission on the Accident at Three Mile Island (Kemeny Report), *The Need for Change: The Legacy of TMI* (Washington, D.C.: GPO, October 1979).

8. Like interview.

9. Kemeny Report, pp. 19, 20. The NRC also inherited the charge derived from the AEC of insensitivity to changing public attitudes about nuclear risk, safety, and the environment (Ebbin and Kaspar, *Citizen Groups and the Nuclear Power Controversy*, p. 224). The remarks about NRC's licensing provisions versus safety are from the Downey interview.

10. NRC chairman Ivan Selin, Interview and Call-In during C-Span television broadast, December 21, 1993.

11. Klein cited in Karl Grossman, *Power Crazy*, p. 280.

12. Uhl interview.

13. See 10 CFR 50.47 (a)(1).

14. Interview with Droggitos, March 3, 1993; 10 CFR sec. 50.47 (c)(1).

15. Interview with William Parler, NRC general counsel, March 22, 1993.

16. Koppelman interview.

17. Suffolk County Resolution No. 262-1982, March 25, 1982; NRC LBP-83-22, 17 NRC 608, April 20, 1983, App. A, "Factual and Procedural Background."

18. Suffolk County Resolution No. 456-1982, May 19, 1982.

19. See *Newsday*, May 31, 1985, for description of Jones; Jones interview.

20. All U.S. commercial power reactors operate inside containments of varying designs intended to prevent uncontrolled release of radioactive materials to the environment. TMI had a "large, dry containment," an immense concrete structure that successfully withstood the overpressure caused by a hydrogen gas explosion outside the reactor pressure vessel. (This was not the "hydrogen bubble" within the pressure vessel that temporarily caused great alarm but, as it turned out, could not have exploded.) The Soviet reactor at Chernobyl had no containment. It is not clear whether all, or even any, of the U.S. containment designs could successfully contain an explosion as powerful as the one that initiated the Chernobyl accident.

21. See 45 FR 55406. As provided in 10 CFR 50.47(c)(2): "the exact size and configuration of the EPZs surrounding a particular nuclear power reactor shall be determined in relation to local emergency response needs and capabilities as they are reflected by such conditions as demography, topography, land characteristics, access routes, and jurisdictional boundaries."

22. *Newsday*, January 2, 1983.

23. See R. J. Budnitz, P. R. Davis, S. Fabic, and H. E. Lambert, "Review and Critique of Previous Probabilistic Accident Assessments for the Shoreham Nuclear Power Station," 2 vols. (Future Resources Associates, Inc., September 17, 1982); F. C. Finlayson and J. H. Johnson Jr., "Basis for Selection of Emergency Planning Zones for the Shoreham Nuclear Power Plant, Suffolk County, New York" (October 1982), draft; J. H. Johnson and Donald J. Zeigler, "Further Analysis and Interpretation of the Shoreham Evacuation Survey," November 1, 1982; PRC Voorhees, "Suffolk County Radiological Emergency Response Plan" (November 1982), working draft report.

24. *Newsday*, January 2, 1983. Within the twenty-mile EPZ, there was a total resident population of about 635,000 people, which was increased by about 100,000 tourists during the summer.

25. Matthew L. Wald, "U.S. Aides Dispute Suffolk on Shoreham Planning," *New York Times*, December 4, 1983.

26. *Newsday*, January 27, 1983; J. H. Johnson and D. J. Zeigler, "A Spatial Analysis of Evacuation Intentions at the Shoreham Nuclear Power Station," in M. J. Pasqualetti and K. D. Pijawka, *Nuclear Power: Assessing and Managing Hazardous Technology* (Boulder, Colo.: Westview Press, 1984), pp. 279–302. See also Social Data Analysts, Inc., "Attitudes Towards Evacuation: Reactions of Long Island Residents to a Possible Accident at the Shoreham Nuclear Power Plant" (June 1982). The findings of the Suffolk poll were described in *Newsday*, July 16, 1982.

27. *New York Times*, April 15, 1979; Thornburgh cited in Transcript of August 21, 1979, Hearing of the President's Commission on the Accident at TMI, John G. Kemeny, Chairman (Washington, D.C.: Bowers Reporting, 1979), p. 21.

28. Social Data Analysts, Inc., "Responses of Emergency Personnel to a Possible Accident at the Shoreham Nuclear Power Plant" (October 1982); Kai Ericson, "Regarding Emergency Planning for the Shoreham Nuclear Power Station," Testimony before the Suffolk County Legislature, January 24, 1983.

29. *Report of the New York State Fact Finding Panel on the Shoreham Nuclear Power Facility* (December 1983), p. 20; *New York Times*, December 3, 1982.

30. *New York Times*, October 7, May 26, 1986.

31. Bordenick interview; *Report of the NYS Fact Finding Panel*, p. 19. See also Suffolk County Legislature, Resolution No. 111-1983, February 17, 1983.

32. *New York Times*, December 10, 1983.

33. Quotes come from interviews with Per Alin, professor of history, State University of New York at Stony Brook, November 7, 1991; and interview with Herbert J. Kouts, chairman, Department of Nuclear Energy at Brookhaven, December 3, 1991.

34. Cordaro interview; Koppelman interview; *Newsday*, January 2, 1983.

35. Cordaro interview; Freilicher interview.

36. Cordaro interview. The *New York Times* reported on January 27, 1980, that Like had contributed $1,250 to the reelection campaign of the former county executive John Klein who ran against Cohalan in the Republican primary in 1979.

37. Quote from *Newsday*, January 2, 1983, April 18, 1982.

38. Cordaro interview; Bordenick interview.

39. *Newsday*, April 18, 1982, May 15, 1988.

40. The characterization of Cohalan comes from interviews with Freilicher and Kouts.

41. Jones interview.

42. Bordenick interview.

43. Kessel interview; Murray interview. (The words of the heading are from the Murray interview.)

44. Koppelman interview.

45. Murray interview; interview with Peter Bradford, former NRC commissioner and former New York PSC chairman, January 9, 1992.

46. *New York Times*, February 18, 20, 1983.

47. *New York Times*, May 8, 1983. See also letter to Robert T. Stafford, chairman, Senate Environment and Public Works Committee, May 16, 1983, p. 3; *Newsday*, June 12, 1983.

48. *Newsday*, June 17, 1983. Simultaneously, in a California case, a ruling issued by the U.S. Supreme Court concluded that "Congress has left sufficient authority in the States to allow the development of nuclear power to be slowed or even stopped for economic reasons" (*Pacific Gas and Electric Co. v. State Energy Resources Conservation and Development Commission et al.*, 461 U.S. 190 [1983]). This decision did not address the federal-state division of regulatory authority as it pertained to Shoreham.

49. Michael Oreskes, "Cuomo Stays Open to Shoreham Plan," *New York Times*, December 20, 1983; editorial, "Mr. Cuomo's Test at Shoreham," *New York Times*, February 23, 1984.

50. John H. Marburger, letter to Governor Cuomo, in *Report of NYS Fact Finding Panel,* December 14, 1983, pp. 2, 1 (words in the heading are from p. 2). See also James Barron, "At Risk," *Empire State Report* (May 1983), p. 38.

51. Letter from John Marburger to the governor, in *Report of NYS Fact Finding Panel,* December 14, 1983, p. 2. The views of panel members are provided as part of the final report of the Marburger Commission.

52. The findings can be found on pp. 29–34 of the *Report*; Murray interview.

53. *Report of the NYS Fact Finding Panel,* pp. 8–37.

54. Murray interview. The quotation in the heading is from *Newsday,* April 18, 1982.

55. *New York Times,* December 15, 1983. According to *Newsday,* December 22, 1983, Dr. Marburger is reported to have said that he personally believed the Shoreham plant could be opened and operated safely.

56. Murray interview.

57. "Quiet and self-contained" is quoted in *Newsday,* January 27, 1983; all other quotes are from *Newsday,* April 21, 1982.

58. *Newsday,* October 6, 1982, February 19, March 9, May 24, 1983. See also *New York Times,* May 28, July 26, 1983.

59. Reporter Stuart Diamond, June 3, 1983. The poll was published in *Newsday,* February 27, 1983. By October 1983, 61 percent of the same group of respondents believed that LILCO should not complete and operate the facility.

60. *New York Times,* June 3, 1983.

61. Ibid., January 16, 1983; *Newsday,* January 20, 1983.

62. *New York Times,* July 3, 1983.

63. Uhl interview; Cordaro interview.

64. Reveley interview. Herbert Brown, quoted in *Newsday,* May 15, 1988. See also *Newsday,* April 18, 1982.

## Chapter 5. Governments in Collision

1. Lindsay Gruson, "NRC Chief Pushed Shoreham Case," *New York Times,* April 17, 1984; Earl Lane, "How LILCO Got U.S. to Listen," *Newsday,* May 7, 1984. At about this time, the nuclear industry was suffering a number of setbacks: Commonwealth Edison's Byron plant was temporarily denied an operating license; Public Service of Indiana abandoned its Marble Hill reactor after spending $2.5 billion; Cincinnati Gas and Electric decided to convert Zimmer

to burn coal; and Philadelphia Electric suspended work on a major nuclear plant of advanced design.

2. Memoranda from Nunzio J. Palladino, NRC chairman, to William J. Dircks, NRC executive director of operations, March 17, 1983, and from Dircks to Palladino, April 1, 1983.

3. Parler interview.

4. LBP-83-22, 17 NRC 608 (April 20, 1983).

5. CLI-83-13, Long Island Lighting Company (Shoreham Nuclear Power Station, Unit 1), 17 NRC 741 (May 12, 1983), and 744 (May 15, 1983).

6. "Revised Emergency Planning Contentions Filed by Suffolk County, Shoreham Opponents Coalition, North Shore Committee Against Thermal and Nuclear Pollution and the Town of Southampton," July 26, 1983, p. 38; "Suffolk County Memorandum in Response to Board Inquiry Regarding Contention 22," August 18, 1983, p. 2 (in NRC HQ PDR) (emphasis in original).

7. LBP 82-19, 15 NRC 601 (March 15, 1982) and 618, 619.

8. ASLB, "Special Prehearing Conference Order," Emergency Planning Proceeding, Docket No. 50-322-OL-3, August 19, 1983, pp. 11 and 9 (in NRC HQ PDR).

9. See CLI 87-12, 26 NRC 383 (November 5, 1987).

10. Downey interview.

11. NRC rules are subject to judicial review within sixty days of the time they are promulgated. After the time for judicial review has passed, a person who believes an NRC rule is not well founded may petition the commission to amend or rescind the rule (10 CFR 2.802). No one filed a timely court challenge to the EPZ rule or petitioned the NRC to change it later on.

12. CLI 87-12, 26 NRC at 393.

13. By excluding the intervenors' contention the commission did not imply that generic rules must be blindly adhered to in licensing proceedings, no matter what. On the contrary, the commission noted that the intervenors could have challenged the safety sufficiency of the ten-mile EPZ if they had shown "special circumstances for the particular site [that is, Shoreham] that were not considered in the emergency planning rulemaking" (26 NRC 395, note 19). Given the unusual geography of the Shoreham site, it would seem that such a showing might have been made. Nevertheless, the commission was correct that the intervenors—litigators with ample legal resources and considerable experience with NRC practice—did not attempt to invoke this exception.

14. Bradford interview.

15. See, for example, "LILCO Transition Plan for Shoreham—Revision 4," Consolidated Regional Assistance Committee (RAC) Review, Attachment 2, October 12, 1984.

16. *Cuomo v. Long Island Lighting Co.*, Consol. Index No. 84-4615, New York Supreme Court, February 20, 1985, pp. 17–18; *Citizens for an Orderly Energy Policy v. County of Suffolk*, 604 F. Supp. 1084 (EDNY 1985).

17. LBP-85-12, 21 NRC 644 (April 17, 1985); LBP-85-31, 22 NRC 410 (August 26, 1985).

18. ALAB-818, 22 NRC 651 (October 18, 1985).

19. LBP-83-21, 17 NRC 593 (April 20, 1983); CLI-83-17, 17 NRC 1032 (June 30, 1983). The wording of the heading is taken from the Murray interview.

20. CLI-83-17 (1983). Gilinsky also expressed his view that "the Commission is in essence playing chicken with the governor of New York" (*New York Times*, December 18, 1983).

21. Until all questions are resolved there is always some uncertainty about whether the plant is going to operate, especially in a hotly contested case, but the likelihood of significant benefits from early low-power testing outweighs the chance that the full-power license will be denied.

22. *Cuomo v. NRC*, 772 F. 2d 972 (D.C. Cir 1985).

23. "The Shoreham Saga," *IEEE Spectrum* (November 1987), p. 32.

24. Murray interview; *Newsday*, August 16, 1983.

25. *New York Times*, February 9, 1984; *Newsday*, December 6, 1983; Lekachman and Luftig, quoted in *New York Times*, April 2, 1984.

26. Quotes from *Newsday*, February 18 and March 13, 1984, and from *New York Times*, November 22, 1984.

27. *Newsday*, January 11, 1984.

28. *New York Times*, May 31, 1985; *Newsday*, May 31, 1985.

29. *Newsday*, January 11, July 17, 1984; *New York Times*, July 17, 1984. By 1996, Catacosinos was earning $580,000 as an annual salary.

30. *IEEE Spectrum* (November 1987), p. 36.

31. Quote from *New York Times*, May 12, 1983; see also *Newsday*, February 9, 22, March 18, 1984; *Washington Post*, May 6, 1984;

32. President Reagan's letter, cited in *New York Times*, October 19, 1984; Keyworth, quoted in *Newsday*, May 7, 1984; DOE meetings, quoted in *Newsday*,

April 23, 1984; Hodel, quoted in *Newsday*, April 24, 1984; Cuomo letter, quoted in *Washington Post*, May 6, 1984; Herrington, quoted in *Inside NRC*, May 13, 1985, and *New York Times*, June 1, 1985.

33. Gilinsky cited in the *Energy Daily*, May 18, 1984; *Newsday*, March 29, 1984. The words of the heading are from Commissioner James K. Asselstine, CLI-84-8, 19 NRC 1154 (May 16, 1984).

34. *Newsday*, April 25, 1984.

35. U.S. Congress, House, Committee on Interior and Insular Affairs, *Hearing, Licensing Process at Shoreham Nuclear Powerplant*, 98th Cong., 2d sess., May 17, 1984, pp. 15–16, 10–13.

36. *New York Times*, June 7, 1984; memorandum, CLI-84-20, 20 NRC 1061 (September 24, 1984); COMJA-84-9, February 5, 1985 (in HQ PDR).

37. The Miller Board took the unusual action of dividing the low-power licensing process into four stages: (1) to load fuel; (2) to operate at an extremely low level (0.001 percent of full power) to test instruments that measure the chain reaction ("cold criticality testing"); (3) testing at 1.5 percent of power ("testing beyond cold criticality"); and (4) testing at 5 percent of power. Phases 3 and 4 were delayed until the diesel generators were found to operate safely ("Victory for LILCO, but Battle Goes On," *New York Times*, November 22, 1984). See also Union of Concerned Scientists, *Safety Second* (Bloomington: Indiana University Press, 1987), pp. 90–92.

38. *Wall Street Journal*, October 30, 1984.

39. Quote from CLI 85-1 21 NRC 275 (February 12, 1985); *Cuomo and Suffolk County v. NRC*, 772 F. 2d 972 (D.C. Cir. 1985), 21 NRC 1587.

40. CLI-85-12 (June 20, 1985). See also *Inside NRC*, June 25, 1984, p. 14.

41. Kessel interview; Cuomo quoted in *New York Times*, July 4, 1985; Palomino interview.

42. Quote from Fred Bernthal, former NRC commissioner, interview, April 19, 1993; Denton interview.

43. Parler interview.

44. Cohalan, quoted in *Newsday*, May 31, 1985; *New York Times*, May 31, 1985. Herrington also said that in a dispute between national policy and local political questions "local politics can [not] be allowed to fly against national security" (Michael Oreskes, "Key Shoreham Foe Halts Opposition to Pullout Drill," *New York Times*, June 1, 1985). The words of the heading are taken from Wayne Prospect, Democratic member of Suffolk County legislature, reported in *New York Times*, June 1, 1985.

45. Executive Order 1, May 30, 1985; "Shoreham: Undamned? Undamned?" *New York Times*, June 5, 1985.

46. *New York Times*, June 1, 1985; *Newsday*, April 25, 1984, June 1, 2, 1985.

47. See, for example, *Matter of Town of Southampton v. Cohalan*, 65 N.Y. 2d 867 and 482 NE 2d 1209 (July 9, 1985).

48. COMTR-85-5A, June 4, 1985 (in HQ PDR). Fabian Palomino described the Shoreham emergency preparedness tests as "Alice in Wonderland exercises" (*Washington Post*, February 14, 1986).

49. Letter from Samuel W. Speck, FEMA, to William J. Dircks, NRC, October 29, 1985; Downey interview; letter from Governor Mario Cuomo to NRC, November 13, 1985; D'Amato, quoted in *Newsday*, January 22, 1986; letter from Senator Patrick Moynihan to Chairman Palladino, February 4, 1986; Jones quote from *Newsday*, May 7, 1984.

50. Local law 2-86, January 16, 1986; *Long Island Lighting Co. and the United States v. County of Suffolk*, 628 F. Supp. 654, 666 (EDNY 1986).

51. Herbert Brown, lawyer for Suffolk County, quoted in *New York Times*, February 13, 1986.

52. LPB-87-32 26 NRC 479 (December 7, 1987).

53. Parler interview.

54. NRC SECY-87-35, February 6, 1987, Attachment A, p. 4 (in HQ PDR). The quote in the heading is taken from Gail Harmon and Ellyn Weiss, letter to NRC commissioners, February 20, 1987.

55. *Inside NRC*, March 16, 1987.

56. Cuomo, quoted in *New York Times*, February 8, 1987, and *Nucleonics Week*, February 12, 1987; Bradford, quoted in *New York Times*, February 7, 1987.

57. Editorial, "Federal Power over Nuclear Power," *New York Times*, February 9, 1987. See also John E. Chubb, "License Seabrook, Shoreham," *New York Times*, August 3, 1987. The *Washington Times* on January 10, 1989, wrote that "a spiraling number of narrowminded state and local officials and their constituents have ignored the unity message of the Founding Fathers in seeking to obstruct nuclear power . . . important to national and international goals."

58. Murray interview; Bordenick interview.

59. Uhl interview; Freilicher interview; Reveley interview.

60. Murray interview; Downey interview.

61. Like interview; Reveley interview.

62. Downey interview; Palomino interview.

63. *Newsday*, May 2, 1986; Palomino interview.

## Chapter 6. Who Decides?

1. Kessel interview.

2. The Suffolk County law was blocked by New York State Supreme Court Justice Weiler as a conflict with the state law, which preempted the LILCO takeover field (*New York Times*, August 8, 1986).

3. *New York Times*, November 24, 1985; *Wall Street Journal*, February 25, 1986. It was also argued that Shoreham's costs could not be passed on to ratepayers unless the plant was operational or "used and useful." Since the state intended to keep Shoreham shut, the stockholders would have to bear the cost of the plant.

4. See, for example, *New York Times*, January 7, 8, 12, 26, June 6, July 3, 12, 1986; *Newsday*, January 27, February 7, 14, 1986. The Sawhill Commission is cited from the *Report of the Long Island Public Power Panel*, June 24, 1986, p. 31.

5. Like interview. However, valuing the stock for friendly or hostile takeover purposes would not have been a simple matter of looking up the numbers in the financial pages of the *Wall Street Journal*. LIPA would have had to offer enough per share to persuade the shareholders to go along with the takeover. The sum the Public Power Panel created by Governor Cuomo thought would be necessary and sufficient was $18 a share, even though the stock was trading for less.

6. "Shoreham Shuffle, Same Old Game," *New York Times*, July 14, 1986.

7. *Long Island Lighting Co. v. Cuomo*, 666 F supp. 370 (NDNY 1987). See also *New York Times*, January 15, 1987, and *Daily News*, August 5, 1987.

8. Josh Barbanel, "Cuomo's LILCO Battle," *New York Times*, July 12, 1986.

9. *Report of the Long Island Public Power Panel*, June 24, 1986, p. 5. The panel pointed out that ratepayer savings "would result exclusively from LIPA's inherent financial advantages" (p. 11). Out of the 7.3 cents per kilowatt hour that LILCO was charging for electricity, 2.2 cents per kilowatt hour went to pay federal income tax.

10. *New York Times*, March 31, 1988.

11. *New York Times*, April 4, 16, 1988.

12. Interview with Joseph W. McDonnell, LILCO vice president, September 20, 1993; Like interview.

13. Freilicher interview; interview with Charles R. Pierce, December 13, 1991; Like interview.

14. See Paul Lewis Joskow, "A Behavioral Theory of Public Utility Regulation," Ph.D. diss., Yale University, 1972, p. 14; Milton Musicus, former energy czar of New York City, quoted in Joan B. Aron, "The 'New Look' at the Public Service Commission," *New York Affairs* 1, no. 4 (1974): 64–65.

15. Aron, "The 'New Look' at the Public Service Commission," p. 71.

16. *Newsday*, November 15, 1981; *New York Times*, December 22, 1985.

17. *New York Times*, May 15, 1986.

18. Kessel interview; Burstein interview. Grumman's chairman cited in *Newsday*, November 21, 1985.

19. Reveley interview.

20. The governor cited in *New York Times*, January 10, 1986. The quotation in the heading is from Philip S. Gutis, *New York Times*, May 29, 1988.

21. Bierwirth interview; Uhl interview; Cordaro interview. The remarks of the CPB appear in New York PSC, "Proceeding on Motion of the Commission as to the Rates and Charges of Long Island Lighting Company for Electric Service," Cases 29484 and 88-E-084, "CPB Initial Brief Supporting Settlement," August 10, 1988, p. 13.

22. Quote from *New York Times*, November 6, 1987; PSC quote from New York PSC, cases 29484 and 88-E-084; "CPB Initial Brief Supporting Settlement," August 10, 1988, p. 2, and *New York Times*, March 30, 1988; plant opponents, quoted in *New York Times*, November 6, 9, 1987; Cuomo, quoted in *Newsday*, November 6, 1987; investment bankers and company officials, quoted in *New York Times*, November 6, 1987.

23. *Wall Street Journal*, January 25, 1988.

24. *New York Times*, May 12, 1988; *Newsday*, May 15, 1988; *New York Journal of Commerce*, May 23, 1988. The *Journal* explained: "The transaction has been structured to entitle LILCO to a $2.5 billion federal tax writeoff. By selling its $2.5 billion plant to the state for $1, the company incurs a loss that will shelter it from federal income taxation for a decade. In other words, LILCO will be nursed back to financial health at federal expense." The announcement of IRS approval was made by LILCO vice president Joseph McDonnell (*New York Times*, September 16, 1988).

25. Like interview.

26. Cuomo, quoted in *Newsday*, May 27, 1988; *New York Times*, May 27, 1988.

27. Wayne Prospect, Long Island legislator, cited in Bill Paul, "Death of

Shoreham Could Help the Nuclear Industry," *Wall Street Journal*, May 27, 1988. Bill Paul suggested that the cost of the New York–LILCO agreement was so high that it could discourage closings of other plants.

28. "Long Island Flexes Its Muscle," *New York Times*, December 2, 1988. Cuomo, quoted in Elizabeth Kolbert, "Legislators Finish Albany Session Without Voting on Shoreham Pact," *New York Times*, December 2, 1988; LILCO cited in Matthew L. Wald, "LILCO Wins Either Way, *New York Times*, December 5, 1988.

29. *New York Times*, September 19, December 3, 1988.

30. *New York Times*, February 12, 1989; Freilicher interview.

31. LILCO vice president, cited in Radio TV Reports, "Nation's Business Today," May 27, 1988; Matthew L. Wald, "LILCO's Competing Voices," *New York Times*, October 5, 1988. Wald also wrote: "Although it has not always been obvious in recent years, utilities prize their relationship with the public."

32. John Rather, "In Albany, Shoreham Confusion," *New York Times*, December 11, 1988.

33. *Newsday*, December 7, 8, 1988.

34. Judge Weinstein ruled that Suffolk's "alleged injuries" were in the form of "utility rate increases" previously approved by the PSC (that would not have been granted save for the misrepresentation) rather than "monetary damages" under RICO (*Southampton Press*, February 13, 1989). See also *New York Times*, June 1, 1986, March 11, 18, September 8, October 4, December 6, 7, 11, 15, 18, 22, 1988, January 19, 29, February 12, 14, 1989, July 5, 1990. See also *Newsday* for a similar period of time.

35. Rex Smith and Susan Benkelman, "Shoreham: It's a Deal," *Newsday*, March 1, 1989.

36. New York PSC, "Opinion No. 89-8," April 13, 1989, pp. 80, 82; Commissioner James A. McFarland, in App. A.

37. *Newsday*, May 28, September 16, 1988, March 1, 1989.

38. Palomino interview; McDonnell interview.

39. Freilicher interview.

40. The secretary of energy is cited in *Newsday*, April 13, June 29, 1989, and in *New York Times*, May 4, 1989; the deputy secretary is cited in *Washington Post*, June 25, 1989; *Wall Street Journal*, June 28, 1989. See also editorials, "The Energy Vandals," *New York Times*, April 20, 1989, and "A Wiser Fate for Shoreham," *New York Times*, March 11, 1989.

41. *Newsday,* March 1, 1989; *New York Times,* March 6, 1990; quotation from Reveley interview.

42. Downey interview.

43. Like cited in William Bunch, ". . . As Shoreham Foes Rejoice Quietly," *Newsday,* June 29, 1989.

44. Bierwirth interview; Burstein interview; Cordaro interview; Freilicher interview.

45. PSC, "Proceeding on Motion of the Commission as to the Rates and Charges of Long Island Lighting Company for Electric Service," Cases 29484 and 88-E-084, "CPB Initial Brief Supporting Settlement," August 10, 1988, p. 13.

46. "Testimony of Dr. Richard A. Rosen," appearing on behalf of the New York CPB Board, Albany, New York, March 22, 1989, Cases 29484 and 88-E-084. Dr. Rosen was an executive vice president of the Energy Systems Research Group, Inc., from Boston, Massachusetts.

47. New York PSC, "Opinion and Order Approving Agreement and Rate Plan," Opinion No. 89-8, April 13, 1989, pp. 8–9, 72.

48. Ibid., pp. 82–83.

49. Peter Bradford, quoted in "The Shoreham War Has Got to End Now," *Newsday,* May 9, 1989; Richard Kessel, quoted in a letter to the editor, *Wall Street Journal,* May 10, 1989. Matthew L. Wald, quoted in "No Clear Winners or Losers Emerging Yet in LILCO Pact," *New York Times,* March 3, 1989. In point of fact, the NRC does not appear to have imposed expensive new requirements on the industry in the five years after Shoreham expired. The trend has been to lower the cost of compliance.

50. Kessel interview.

51. Bradford interview.

52. Downey interview. The words of the heading are from Kessel interview.

53. Sailor interview; Wald interview (by telephone); Bordenick interview.

54. Reveley interview.

55. Kessel interview.

56. Murray interview.

57. Uhl interview; Freilicher interview; Ben Wattenberg, "Nuclear Power: Cuomo vs. the Feds," *New York Post,* July 25, 1989. *New York Times* reporter Elizabeth Kolbert wrote that Cuomo "finally acted to accomplish what he said he believed in" (March 1, 1989).

58. Murray interview; Like interview.

59. Editorial, "Lights Out," *Wall Street Journal,* May 31, 1988; Bierwirth interview; *New York Times,* May 13, 1988. See, for example, editorials titled "The Pied Piper of Shoreham," January 24, 1989; "Bobbing and Weaving on Shoreham," February 17, 1989; and "Fear of Shoreham," March 2, 1989. Similarly critical views surfaced in the Kouts interview and the interview with Dr. Miro M. Todorovich, executive director, Scientists and Engineers for Secure Energy, Inc., November 8, 1991; quote from Todorovich.

60. Reveley interview.

61. Herbert Brown, "CPB Reply Brief Supporting Settlement," Cases 29484 and 88-E-084, August 16, 1988, appendix A.

### Chapter 7. Dead on Arrival

1. Executive Order No. 12657, November 18, 1988. The words of the heading are taken from Marilyn Goldstein, *Newsday,* March 3, 1989.

2. President Reagan's earlier statement was made in a letter to then Representative Carney, who was waging a fierce battle for reelection (*New York Times,* October 19, 1984). White House officials, quoted in Wines, "The Nuclear Power War," *New York Times,* November 21, 1988; McDonnell, quoted in Philip S. Gutis, "Reagan Directive Is Called Too Late to Save Two Plants," *New York Times,* November 20, 1988; "Mr. Reagan and the Reactors," *Washington Post,* November 25, 1988.

3. Letter from Herrington to NRC chairman, Lando Zech Jr. The assertion that this was a "prohibited ex parte" communication was made in a letter from William Parler, NRC general counsel, to Herrington. Both letters bear the date October 7, 1988.

4. Letter from Zech to Congressman Hochbrueckner, August 18, 1988. See also *Inside NRC,* August 29, 1988.

5. *New York Times,* February 3, 1988; letter from Grant C. Peterson, associate director, FEMA State and Local Programs and Support, to Victor Stello Jr., NRC executive director for operations, May 31, 1988; letter from Zech to Julius W. Becton Jr., FEMA director, June 3, 1988; and letter from Peterson to Stello, September 9, 1988.

6. The state and county said: "We would never follow LILCO's plan or coordinate in any way with LILCO" (see 28 NRC at 358). Gleason board cited from LBP-88-24, 28 NRC 311 (September 23, 1988); *New York Times,* September 24, 1988.

7. See 28 NRC at 391.

8. At this time, another board was considering matters in connection with an emergency preparedness exercise recently conducted at Shoreham (28 NRC 423, October 7, 1988).

9. ALAB-902, 28 NRC 423 (October 7, 1988); Philip S. Gutis, "Limited License for Shoreham Backed," *New York Times*, November 22, 1988; Murray interview.

10. CLI-89-02, 29 NRC 211 (March 3, 1989); the words of the heading are taken from Koppelman interview.

11. Murray interview; Like interview; Bradford interview.

12. "Director's Finding Regarding Shoreham Emergency Preparedness," April 17, 1989; *New York Times*, March 4, April 21, 1989.

13. Observers quoted in *New York Times*, April 4, 1989; Clifford D. May, "Shoreham Gains Full License, Despite Plan to Scrap It," *New York Times*, April 21, 1989.

14. CLI-89-02, 29 NRC 211 (February 28, 1989).

15. "Too Little, Way Too Late," *New York Daily News*, April 23, 1989.

16. Zech, quoted in Clifford D. May, "Shoreham Gains Full Power License," *New York Times*, April 21, 1989. Heading comes from Bierwirth interview.

17. LILCO's Comment on the Immediate Effectiveness of LBP-88-24, October 3, 1988; *Newsday*, April 23, 1987, April 18, 1989. See also *Inside NRC*, August 29, 1988; CLI-89-02.

18. Cuomo and Watkins, quoted in *New York Times*, April 21, 1989; DOE deputy secretary, quoted in *New York Times*, April 22, 1989; Watkins, quoted in *New York Times*, April 28, 1989. See also Dirk Victor, "The Nuclear Turn-On," *National Journal*, September 9, 1989.

19. *Inside NRC*, November 20, 1989.

20. Sununu, quoted in Kinsey Wilson, "Cuomo Wants Meeting on N-Plant," *Newsday*, July 18, 1989; Cuomo, quoted in Thomas W. Lippman, "Cuomo, DOE Aide Spar over Shoreham," *Washington Post*, November 10, 1989; LILCO, quoted in Matthew L. Wald, "Competing Designs on Shoreham Remain," *New York Times*, May 7, 1989; and *Newsday*, July 18, 1989.

21. Bierwirth interview. Heading is taken from Secretary James Watkins, quoted in Clifford D. May, "U.S. Energy Chief Vows to Fight Shoreham Closing," *New York Times*, April 28, 1989.

22. See Matthew L. Wald, "Competing Designs on Shoreham's Remain," *New York Times*, May 7, 1989; Philip S. Gutis, "Delay and Shoreham," *New York*

*Times*, July 31, 1989; and letter from Richard Wilson to *Newark Star Ledger*, June 1, 1989, in re. the "politics of delay." Descriptions of plant supporters come from *New York Times*, July 28, 1989, July 31, November 12, 1991, and *Newsday*, June 6, 1990.

23. *Newsday*, May 3, 1989; *New York Times*, May 7, 1989; *Suffolk Life*, May 10, 1989.

24. DOE request from *New York Times*, May 7, 1989; *Inside Energy*, June 19, 1989; Flynn, quoted in *Inside NRC*, June 19, 1989.

25. Rose Gutfeld, "Bush Administration Joins Suit," *Wall Street Journal*, July 14, 1989; DOE, quoted in Philip S. Gutis, "Federal Plan to Save Shoreham Unveiled," *New York Times*, July 29, 1989; Kinsey Wilson, "U.S. Joins Shoreham Boosters," *Newsday*, September 8, 1989; "Judge Rejects U.S. Try," *New York Tribune*, October 6, 1989.

26. Alin interview.

27. See *Cuomo v. NRC*, 772 F. 2d 972 (D.C. Cir. 1985).

28. Earl Lane in *Newsday*, May 4, 1989.

29. Matthew L. Wald, "U.S. Approves Decommissioning of Unwanted Shoreham A-Plant," *New York Times*, June 13, 1992.

30. Kinsey Wilson, "Nail in Shoreham Coffin," *Newsday*, April 6, 1990; *Inside NRC*, April 9, 23, 1990.

31. ". . . and Shoreham Can Still Make a Difference," *New York Post*, August 9, 1990; "Parochial Politics on Oil," *New York Times*, August 21, 1990; John O. Sillin, "The Nuclear Energy Weapon," *Wall Street Journal*, August 23, 1990; "Iraq Could Revive Nuclear Power in U.S.," *Christian Science Monitor*, August 24, 1990.

32. Kinsey Wilson, "Bush Panel Joins Shoreham Fray," *Newsday*, October 17, 1990; *Inside NRC*, October 8, 1990; "Amicus Submission by DOE," transmitted in a letter to Admiral Kenneth M. Carr, NRC chairman, November 9, 1990; PSC chairman, quoted in *Inside Energy*, December 3, 1990.

33. Judge Williams, quoting LILCO, in *Shoreham–Wading River Central School District v. NRC*, 931 F. 2d 102, 107 (D.C. Cir. 1991); *Newsday*, June 6, 1990. The heading is taken from Thomas A. Twomey Jr., cited in Matthew L. Wald, "NRC Modifies Stand," *New York Times*, November 4, 1990.

34. *Inside NRC*, October 22, 1990; Matthew L. Wald, "NRC Modifies Stand," *New York Times*, November 4, 1990.

35. There is judicial authority contradicting the commission's argument. In *Natural Resources Defense Council v. Morton*, 458 F. 2d 827 (1972), the U.S.

Court of Appeals for the District of Columbia held that the alternatives an agency must consider are not limited to measures the agency has authority to adopt. The impact statement is intended to be of use to other decision makers besides the agency itself, including Congress and the president (ibid., 834).

36. *New York Times*, November 4, 1990.

37. LIPA, "Supplement to Environmental Report (Decommissioning)," Shoreham Nuclear Power Station, NRC Docket No. 50-322, (December 1990), pp. 1–4 (in HQ PDR).

38. Heading comes from Sarah Lyall, "Court Says Shoreham A-Plant Can Be Closed," *New York Times*, October 23, 1991.

39. Watkins, quoted in *Inside Energy*, January 28, 1991. School District and SE2 had argued unsuccessfully that the utility had to submit a decommissioning plan before a POL was issued and that the "decommissioning report" already submitted to NRC fell short of a plan (CLI 91-1, 33 NRC 1 [January 24, 1991]). See also *Inside NRC*, January 28, 1991; and *Inside Energy*, January 28, 1991. Once more DOE bemoaned the "adverse consequences of this decision [to destroy the plant] on energy security and reliability, the environment, the ratepayers of Long Island and the New York State economy" (National Association of Regulatory Utility Commissioners, *Bulletin*, February 4, 1991).

40. Sarah Lyall, "NRC Clears the Way," *New York Times*, June 13, 1991; Kinsey Wilson, "Losing Its License," *Newsday*, June 13, 1991.

41. Sarah Lyall, "U.S. Sues for an Environmental Study," *New York Times*, July 12, 28, 1991; *Inside NRC*, July 15, 1991. (The action of the U.S. Court of Appeals and the Supreme Court took place on July 19 and August 2, 1991.)

42. *New York Times*, October 23, 1991; Bradford, quoted in *Washington Post*, June 25, 1989.

43. Thomas W. Lippman, "Nearly New A-Plant Going to Scrap Heap," *Washington Post*, June 18, 1992.

44. *Inside NRC*, June 15, 1992. The earlier incident is described in the *New York Times*, May 31, 1985, and *Newsday*, May 31, 1985.

45. Kinsey Wilson, "Shoreham Fuel Going Upstate?" *Newsday*, January 4, 1991, October 13, 1994. See also Matthew L. Wald, "New York's Nuclear Fuel May Be Headed Abroad," *New York Times*, September 23, 1992.

46. *State of New Jersey Department of Environmental Protection and Energy v. LIPA, NRC, U.S. Coast Guard, and Philadelphia Electric Co.*, 30 F. 3d 403 (3d Cir. 1994).

47. Like and Freeman, quoted in "Uranium Shipment Ends Shoreham Reactor's Nuclear Life," *New York Times,* June 4, 1994.

48. Letter from Clayton L. Pittiglio Jr., project manager, to Mr. Frederick Petschauer, resident manager, LIPA, April 11, 1995, enclosing an Order Terminating the Shoreham Nuclear Power Station, Unit 1, Facility Operating License No. NPF-82; Jones, quoted in *Newsday,* May 2, 1995.

49. *Newsday,* January 4, July 12, 1991; *New York Times,* March 19, May 4, 1989. See also letter from LILCO president Anthony F. Earley Jr. to NRC, January 27, 1994. In 1994, LILCO estimated the total cost of decommissioning, including fuel transfer, to be about $500 million, but this total was exclusive of all taxes and LILCO carrying costs.

50. "A Wiser Fate for Shoreham," *New York Times,* March 11, 1989; Karl Grossman, *Southampton Press,* April, 13, 1989; Adam Z. Horvath, "Proposal," *Newsday,* September 13, 1989; John T. McQuiston, "Shoreham Is Proposed as Terminal," *New York Times,* March 29, 1993.

51. *New York Times,* June 23, 1991, August 20, 1993, May 22, 1994.

52. Murray interview.

53. Charles Pierce, LILCO chairman, quoted in James Barron, "At Risk," *Empire State Report,* May 1983, p. 36. "State and LILCO in Clash on Power Outlook for Long Island," *New York Times,* December 29, 1985. See also "Two Utility Groups Back Shoreham," *Newsday,* February 19, 1986; *New York Journal of Commerce,* June 16, 1986. The Marburger Commission also found that the projections for Long Island's future electrical energy needs on which Shoreham was based were obviously "overestimates." LILCO had enough generating capacity to satisfy probable demand until at least 1993 and probably longer (*Report,* p. 37). The statement of Governor Cuomo's spokesman appeared in "State Study on Energy Assailed by LILCO," *New York Times,* May 24, 1987. See also North American Electric Reliability Council's "1988 Summer Assessment" of overall reliability of electric supply, which found that "generation capacity will be marginal" due to the continued delay in Shoreham's operation ("Brownouts and Blackouts?" *Barron's Weekly,* May 30, 1988).

54. *New York Times,* April 2, May 28, 1988; *Newsday,* April 2, 1988. See also *New York Times,* April 17, 21, 1988, May 5, 6, July 9, 23, 1989.

55. Peter Bradford, "Changing Regulation to Further the Goals of Energy Efficiency," 1989 Competitek Forum, Rocky Mountain Institute, Snowmass, Colorado, September 26, 1989; Murray interview.

56. Murray interview.

57. Bradford interview; Uhl interview.

## Chapter 8. The Message

1. Statement of Representative Joe Barton (R-Texas), quoted in *Nucleonics Week*, May 21, 1992, p. 1.

2. See 10 CFR 52.103.

3. See, for example, the remarks of Dixie Lee Ray, former head of the AEC, who said that citizens could have valid opinions in "social" decisions but "with a scientific and technical decision (such as nuclear energy) this is not true" (*Weekly Energy Report*, no. 1 [1973]: p. 7). Although these remarks were made more than twenty years ago, they still reflect the prevailing view within the nuclear industry and the NRC.

4. The united pronuclear front of prominent scientists fell apart years ago when Henry Kendall, a professor of physics who later received the Nobel prize, founded the Union of Concerned Scientists, which has brought many challenges, in and out of court, to what it deemed were inadequate or unduly confident assessments of safety issues affecting nuclear power.

5. Those who support public participation in the decision-making process believe, with Alvin Weinberg and others, that "nuclear technology can coexist with open, participatory, decentralized democracy" (letter on "Splitting the Atom, Not the Country," *Wilson Quarterly* [New Year's, 1986]: p. 173). See also Brian Balogh, *Chain Reaction: Expert Debate and Public Participation in American Commercial Nuclear Power, 1945–1975* (Cambridge: Cambridge University Press, 1991); and Walter A. Rosenbaum, "What Public Participation Can and Can't Accomplish: Notes on the Environmental Experience" (Woodrow Wilson International Center for Scholars, Smithsonian Institution, Washington D.C., June 12, 1979), among many others.

6. Uhl interview.

7. Malsch interview; Reveley interview. Reveley also points out that, in his experience in the nuclear area, "every time a small activist group cares passionately about something, they never think the process is fair unless, on the merits, it decides their way."

8. Selin, quoted in Thomas W. Lippman, "NRC Chief Seeks to Restore Nuclear Power's Image and Fortunes," *Washington Post*, December 2, 1991.

9. Chairman Selin's remarks were made in his opening statement to the Sen-

ate in his confirmation hearings and were distributed to all NRC employees on July 9, 1991. Dr. Miro N. Todorovich, executive director of SE2, would agree with this recommendation. He thought that the NRC's visibility was "almost nil" and suggested that NRC should "make more visible how thoroughly and seriously they pursue their business and explain how carefully they do their work at a level that's never been done before" (interview, November 8, 1991).

10. Admiral James Watkins to the Nuclear Power Assembly, May 27, 1990. Watkins also said, "Scientifically, in many ways, the uneducated American has been bamboozled and hyped by the anti-nuclear movement." See also remarks of former NRC Chairman Carr, who believes the NRC "didn't do a good job of educating the public" (*New York Journal of Commerce*, July 24, 1989).

11. Ahearne's remarks are found in *Inside NRC*, January 1, 1990, and *Improving Risk Communication* [Washington, D.C.: National Academy Press, 1989), p. 9.; U.S. Congress, Office of Technology Assessment, *Nuclear Power in an Age of Uncertainty*, OTA-E-216 (Washington, D.C.: GPO, February 1984), p. 25. Finger's remarks appear in *Issues in Science and Technology* (spring 1990), p. 8. Open dialogue with the public may not always accord with the commission's views either. *Time* did a cover story in 1996 dealing with whistle-blower complaints about nuclear safety concerns and rule violations at the Northeast Utilities Millstone plant in Connecticut and observed: "Though the NRC's Mission Statement promises full accountability—'nuclear regulation is the public's business' it says—the agency's top officials at first refused to be interviewed" (Eric Pooley, "Nuclear Warriors," Special Investigation: Blowing the Whistle on Nuclear Safety, March 4, 1996, p. 48). The agency's top officials, James Taylor, executive director of operations, and his deputy have since retired from the NRC.

12. Reveley interview. Sawhill is quoted in Bill Paul, "Demise of Shoreham . . . ," *Wall Street Journal*, April 21, 1989; editorial, "Federal Power over Nuclear Power," *New York Times*, February 9, 1987; Howard Shapar, "Nuclear Power: Technical and Institutional Options for the Future" (Washington, D.C.: National Academy Press, 1992), p. 201. Peter Bradford, on the other hand, calls federal preemption a "tragic mistake" (*Wall Street Journal*, February 6, 1987).

13. Denton interview. Note that Martin Malsch, former deputy NRC general counsel, told a commission of Massachusetts lawmakers that oversight of nuclear plant operation may be "a valid exercise of state power" (*New York Journal of Commerce*, February 4, 1987). See also John L. Campbell, who asks, if we

chose to go the route of government intervention to save the commercial nuclear option, "whose interest will be served?" and what impact will it have on our democratic processes? (*Collapse of the Industry: Nuclear Power and the Contradictions of U.S. Policy* [Ithaca: Cornell University Press, 1988], p. 196).

14. Reveley interview; Parler interview. Parler may well be correct. The commission's "realism" rule in principle would thwart a state's attempt to use noncooperation in emergency planning to stop a plant from being licensed, but even the "realism" rule requires that an adequate emergency response is possible and can be implemented.

15. See the National Academy of Sciences report on "Policy Implications of Greenhouse Warming," described in *Inside NRC*, April 22, 1991, p. 6; *Nuclear Power: Technical and Institutional Options for the Future* (Washington D.C.: National Academy Press, 1992), as described in *Inside Energy*, June 22, 1992, p. 8; Alvin M. Weinberg, *The First Nuclear Era*, p. 279.

16. Admiral Zech's remarks were made in a speech to the Southeastern Electric Exchange, Boca Raton, Florida, March 24, 1986; Carr, quoted in Matthew L. Wald, "Nuclear Agency's Chief Praises Watchdog Groups," *New York Times*, June 23, 1992; Selin, quoted in Thomas W. Lippman, "NRC Chief Seeks to Restore Nuclear Power's Image," *Washington Post*, December 2, 1991.

17. The NRC's moves toward "public responsiveness" are described in *Inside NRC*, June 15, 1992, p. 9, January 9, 1995, p. 1, and January 23, 1995, p. 1, and in *NR&C*, February 1995. The Millstone events are described in *Time*, March 4, 1996, pp. 47–54, and in newspaper articles.

18. Reveley interview. He also believes that Shoreham has made more nuclear law than any other plant in establishing the principle that "if you [a utility] decide that you want to get out, as LILCO ultimately did, you will be allowed to get out."

## Epilogue

1. The wording of the subheading is from John J. LaMura, Brookhaven supervisor, *New York Times*, Long Island edition, August 26, 1994.

2. Kinsey Wilson, "It's More Than Meets the Eye," *Newsday*, April 2, 1991; "World Heavy-Rate Contender," July 14, 1991; *New York Journal of Commerce*, July 5, 1991; Jonathan Rabinovitz, "State Accuses LILCO of Deception," *New York Times*, March 1, 1994. One can only speculate whether LILCO's fuel costs might have been reduced or avoided if Shoreham had operated.

3. John Rather, "LILCO Not a Typical Case," *New York Times*, October 10, 1995.

4. Wall Street reaction described in Molly Baker, "State Plan to Buy LILCO Meets with Wariness," *Wall Street Journal*, October 17, 1994; editorial, "The Governor's Plan for LILCO," *New York Times*, October 20, 1994. The *Times* thought: "LILCO could use the low-cost loans to buy back shares, thereby eliminating dividends, and replace existing high-cost loans." See also *New York Times*, October 14, 15, December 19, 1994.

5. Peter Marks, "LIPA Makes $9.2 Billion Offer," *New York Times*, June 21, 1995; James Dao, "Pataki Leans to Plan He Once Scorned," *New York Times*, September 12, 1995. See also Alan Finder, "Pataki Seeking LILCO Breakup," September 12, 1995.

6. "After Twenty Years, a Day of Reckoning on Shoreham," *Newsday*, November 17, 1996; Bruce Lambert, "An Electric Shock," *New York Times*, November 9, 1996.

7. Gaffney and Brand, quoted in "LILCO Gets Upper Hand," *Newsday*, November 10, 1996; Like, quoted in Bruce Lambert, "Agency Votes to Start Talks," *New York Times*, February 29, 1996.

8. "The LILCO Merger Poses Problems," letter to *New York Times*, January 19, 1997.

9. John Rather, "LILCO Case Returns to Power Authority," *New York Times*, January 12, 1997.

# Index

1154

# DATE DUE

| | | | |
|---|---|---|---|
| | | | |
| | | | |
| | | | |
| | | | |
| | | | |
| | | | |
| | | | |
| | | | |
| | | | |
| | | | |
| | | | |
| | | | |
| | | | |
| | | | |
| | | | |
| | | | |
| | | | |

DISCARDED

DEMCO 38-296